DESIGNING AND EXECUTING STRATEGY IN AVIATION MANAGEMENT

*Dedicated to our beloved parents,
Georgios and Vasiliki Flouris and Edward and Helen Oswald*

Designing and Executing Strategy in Aviation Management

TRIANT G. FLOURIS
San Jose State University, USA

and

SHARON L. OSWALD
Auburn University, USA

ASHGATE

© Triant G. Flouris and Sharon L. Oswald 2006

All rights reserved. No part of this publication may be reproduced, stored in a retrieval system or transmitted in any form or by any means, electronic, mechanical, photocopying, recording or otherwise without the prior permission of the publisher.

Triant G. Flouris and Sharon L. Oswald have asserted their moral right under the Copyright, Designs and Patents Act, 1988, to be identified as the authors of this work.

Published by
Ashgate Publishing Limited
Wey Court East
Union Street
Farnham
Surrey GU9 7PT
England

Ashgate Publishing Company
Suite 420
101 Cherry Street
Burlington, VT 05401-4405
USA

Ashgate website: http://www.ashgate.com

British Library Cataloguing in Publication Data
Flouris, Triant G.
  Designing and executing strategy in aviation management
  1.Airlines - Management 2.Strategic planning
  I.Title II.Oswald, Sharon L.
  387.7'068

Library of Congress Cataloging-in-Publication Data
Flouris, Triant G.
  Designing and executing strategy in aviation management / by Triant G. Flouris and Sharon L. Oswald.-- 1st ed.
       p. cm.
  Includes bibliographical references and index.
  ISBN 0-7546-3618-6
 1. Airlines--Management. 2. Strategic planning. I. Oswald, Sharon L. II. Title.

HE9776.F57 2006
387.7068'4--dc22

2005028246

ISBN-978-0-7546-3618-2

Reprinted 2007, 2009

Mixed Sources
Product group from well-managed forests and other controlled sources
www.fsc.org Cert no. SA-COC-1565
© 1996 Forest Stewardship Council

Printed and bound in Great Britain by
MPG Books Ltd, Bodmin, Cornwall.

Contents

List of Figures	x
List of Tables	xi
List of Boxes	xii
Foreword	xiv
Preface	xvi
Acknowledgements	xx
List of Abbreviations	xxi

1 The Essence of Strategy — 1

What is Strategic Management?	1
What is the Strategy of an Organization?	2
The Meaning of Competitive Strategies	3
The Difference between Operational Effectiveness and Strategy	4
The Strategic Management Process	5
Corporate Values and Social Responsibility	6
Designing a Corporate Vision	8
Is a Strategic Vision Really Important?	10
Sample Vision Statements	10
Designing a Mission Statement	11
The Reason Mission Statements are Important	12
Mission Statements and the Environment	12
Setting Organizational Objectives	13
Long-Range and Short-Range Objectives	14
Objectives Should be Pervasive Throughout the Organization	14
Strategic Versus Financial Objectives	14
Carving Out a Strategy	15
Strategy is an On-going Process	16
Sample Mission Statements	16

2 Strategic Positioning and Sustaining a Market Presence — 19

The Generic Strategies	19
Low-Cost Leadership Position	21
When Should a Low-cost Leadership Strategy be Used?	24
What do Managers Have to Do to Achieve Low-Cost Leadership?	24

Differentiation Strategies ... 25
Niche Strategy ... 29
   Geographical Niche ... 29
   Customer-Type Niche ... 29
   Product-line Niche ... 30
   Cost or Differentiated Niche Strategy ... 31
Best-Cost Producer Strategy ... 34
The Miles and Snow Typology ... 34
   Defender ... 35
   Prospector ... 35
   Analyzer ... 35
   Reactor ... 35
A Fresh Perspective on Competitive Strategies ... 36
A Summary of Competitive Strategy ... 37

3 The Essence of Competitive Strategies ... 39

Flexibility and Competitive Advantage ... 40
Core Competency ... 42
How Does One Recognize a Core Competency? ... 42
   Are Our Skills Truly Superior? ... 42
   How Sustainable is the Superiority? ... 43
   How Much Value Can the Competency Generate in Comparison to
      Other Economic Levers? ... 44
   Is the Competence Integral to Our Value Proposition? ... 44
Turning Core Competencies into Sustainable Competitive Advantages ... 44
   Sustainable Competitive Strategies ... 44
   Fit and Focus ... 45
Can any Strategic Position be Copied? ... 45
   When are Tradeoffs Important? ... 46
   Tradeoffs are Essential to Sustainability ... 46
Case Illustration: Institutionalizing Competitive Advantage: Southwest
   Airlines' Unique Advantage ... 47

4 The External Environment ... 51

The Macro External Environment ... 52
The Micro External Environment ... 55
Michael Porter's Five Competitive Forces ... 56
   Barriers to Entry ... 57
   The Power of the Supplier Market ... 58
   The Power of the Buyer Market ... 59

The Power of Substitute Products	60
Rivalry among Industry Firms	61
Industry Characteristics	62
Understanding Opportunities and Threats in the Industry	63
Driving Forces	65
What is a Driving Force?	65
Understanding the Competition	66
Key Success Factors	69
The Process of Environmental Analysis	70
Environmental Scanning	70
Monitoring	71
Forecasting Environmental Change	71
Assessing Environmental Change	71
Using Experts to Help in Environmental Assessment	72
The Delphi Technique	73
Organizational Brainstorming	74
Tools to Address Environmental Uncertainty	74
Scenario Planning	74
The Limitations of Environmental Analysis	75
Case Illustration: The Denver International Airport: An Environmental Debacle	76

5 The Internal Environment — 79

Value Chain Analysis	80
How Does One use the Value Chain?	81
What are the Problems with Value Chain?	88
Strategic Cost Analysis	88
Resource-Based View	89
Strategy in the Twenty-first Century	90
Benchmarking: Learning from Others	90
Focusing on the Customer	91
Outsourcing	91
Strategy and the Internet	93
Knowledge Management	94
Case Illustration: JetBlue: Value Added	95

6 Setting Corporate Direction — 99

What Will it be – Single or Multi-Business?	99
Corporate Strategies	100

Growth Strategies 101
   Intensive Growth Strategies 101
   Integrative Growth Strategies 103
   Concentration Growth Strategies 105
   Diversification Growth Strategies 105
No Growth Strategies 111
   Retrenchment and Turnaround 111
   What are the Common Themes to Turnaround? 113
   Liquidation 113
   Divestiture 114
International Strategies 115
   Strategies Tailored to Specific Situations 116
Strategies for Fragmented Industries 116
Strategies for Declining Markets 117
Strategies for Emerging Industries 118
   First-Mover Advantages 118
Case Illustration: AA's Acquisition of TWA: Timeline of Events 119

7   Establishing a Strategy 125

SWOT Analysis 125
   Strengths 126
   Weaknesses 126
   Opportunities 127
   Threats 128
   What else Can be Gained from SWOT? 132
   Formulating Strategy under Uncertainty 132
   Choices of Responses to Uncertainty 135
Case Illustration: Background to Air Transport Regulation and Attempts for Liberalization 137

8   Aviation Strategy Implementation 143

Strategic Leadership: A Key to Successful Implementation 145
   Preparation 145
   Leadership 146
   Change 146
   Partnership 146
Knowing When to Hold it and When to Outsource 146
Putting Together the Right Staff 148
Matching the Right Organizational Structure to a Strategy 150
   The Simple Structure 151

The Functional Structure	151
Multidivisional Structure	152
Strategic Business Units	156
The Matrix Organizational Structure	157
Instituting Total Quality	161

9   Managing Strategy Execution through Tracking, Support Systems and Controls — 165

Tracking through Information Systems	165
Internal Systems	166
E-Commerce as a Support System	167
Other Essential Support Systems	170
Employee-level Controls	171
Corporate Governance Controls	173
Strategic Controls and Strategic Change	175
Implementation Controls	176
Bibliography	*177*
Index	*181*

List of Figures

4.1	The Macro External Environment	52
4.2	The Micro Environment	55
4.3	Porter's Five Forces of Competition	56
4.4	Examples of Several U.S. Carriers in Terms of Geographic Scope and Price	67
4.5	Examples of Several Non-U.S. Carriers in Terms of Geographic Scope and Price	68
4.6	Assessing Environmental Change through Plotting Trends	72
5.1	Value Chain	81
5.2	Core Processes in a Passenger Airline Business Model	84
6.1	Related Diversification	106
7.1	SWOT Analysis	125
8.1	Simple Organizational Structure	151
8.2	Functional Organizational Structure	152
8.3	Southwest Airlines' Modified Functional Organizational Structure	153
8.4	Multidivisional Geographic Organizational Structure	154
8.5	Product-line Multidivisional Structure for an Airline	154
8.6	Multidivisional Business Organized in SBUs	159
8.7	Matrix Organizational Structure	159
8.8	Lines of Authority in Matrix Organizational Structure	160

List of Tables

1.1	Cost Per Available Seat Mile in 2002	4
2.1	Comparative Schedules and Prices between Ryanair, SAS and British Airways	23
2.2	Chalk's Timetable	31
3.1	Historical Operational and Service Offerings	40
3.2	New Operational and Service Offerings	41
4.1	Opportunities for the Aviation Industry Post-September 11	63
4.2	Threats to the Aviation Industry Post-September 11	64
4.3	Competitive Matrix for the Airline Industry	70
5.1	Value Chain through an Integrated Software Architecture	86
5.2	Examples in an Airline where Business Process Outsourcing is Applicable	92
6.1	American Airlines Corporate Information Pre- and Post-TWA-Takeover	123
7.1	Jet Airways' SWOT Analysis	130
8.1	The Divisions that Make up Lufthansa's SBU Organizational Structures	158
8.2	2004 Airline Quality Rating	163

List of Boxes

Ch. 1	Values, Vision, Mission	7
	Vision Statements	10
	What Questions Should be Answered by the Mission Statement?	12
	Financial vs. Strategic Objectives for the Aviation Industry	15
Ch. 2	Low-Cost Leadership	20
	What are the Competitive Strengths of a Low-Cost Leadership Strategy?	22
	When a Low-Cost Strategy Works Best	24
	What are the Problems with a Low-Cost Leadership Position?	26
	JetBlue Airways' Differentiation Strategy Approach	27
	What are the Competitive Strengths of a Differentiation Strategy?	27
	When a Differentiation Strategy Works Best	28
	What are the Problems with a Differentiation Position?	28
	British Airways Concorde Service	29
	What are the Competitive Strengths of a Niche Strategy?	32
	When a Focus Strategy Works Best	32
	What are the Problems with a Niche Position?	33
	Adapting the Miles and Snow Typology to the Regional Airlines	37
Ch. 3	Questions to Ask to Determine Core Competencies:	43
Ch. 4	Eclipse Jet and Friction Stir Welding7	54
	When to Use Scenario Planning:	75
Ch. 5	Defining the Value Chain	82
	Support Activities5	82
Ch. 6	United Parcel Service: Related Diversification	107
	US Airways' Woes	112
	Dobbs International Sold to SwissAir Parent	115
	Characteristics of Fragmented Industries	116
Ch. 7	Sources of Strengths	126
	Sources of Weakness	127
	Sources of Opportunities	128
	Sources of Threats	128
Ch. 8	Denver Airport Revisited	145
	Outsourcing Aircraft Maintenance	147
	The Case of United Airlines	149
	JetBlue's Organizational Behavior	150
	Delta and Song	155

	Lufthansa's SBUs	156
	Boeing: The Matrix Organization	158
Ch. 9	E-Commerce and the Airlines	169
	United Airlines' Statement on Corporate Governance	175

Foreword

When we started SkyEurope in September 2001 we aimed at creating the first "low-fare, low-cost, no-frills" airline based in the heart of Europe in order to provide expanded travel options to both leisure and business customers. The cornerstone of our business model is the stimulation of "point-to-point" service from cities within our network. We believe in serving our customers through the provision of safe, affordable, and friendly service, and we have based our operations on this philosophy. We have also sought to create a friendly and motivating work environment that supports our employees in performing their jobs. Currently, SkyEurope is one of Europe's fastest growing low-cost passenger airlines using our four Central European bases, Bratislava, Budapest, Warsaw, and Krakow, as pivots in providing direct airline travel options to underserved markets throughout Europe.

The successful implementation of our business model has allowed us to become the largest low-cost, low-fare carrier based in Central Europe based on our current fleet size, which has a total capacity of 1,676 seats as of 1 December 2005. We believe that our business will continue to grow as we have "placed" our business in a significant population hub. The three countries in which our bases are located, the Slovak Republic, Hungary, and Poland, collectively comprise more than two-thirds of the total population of the countries that acceded to the European Union in May 2004, which we believe provides significant growth opportunities for us. In addition, our largest base, Bratislava, where we also maintain our operational headquarters, is located approximately 50 kilometres from Vienna, thereby providing us with access to Austria, one of the most mature air travel markets in Central Europe.

Today, SkyEurope operates a leased fleet of 12 Boeing 737 (Classics) aircraft. We have also recently ordered 16 brand new Boeing 737-700 (Next Generation) and have acquired purchase rights for an additional 16 Boeing 737-700 (Next Generation) aircraft. Our aim is to create a single-aircraft-type fleet so as to maximize operational efficiency and minimize aircraft utilization costs. We want to continue to be a leading low-cost operator and to pass on these low costs to our customers.

However, we still have a lot of work to do in solidifying our position in the very competitive intra-European airline markets, and through clear focus on our business model and continuous innovation I believe we will continue to be successful in bringing safe and affordable airline travel options to our customers. Running a low-cost airline, or any airline, especially in a very competitive market is a perilous task for those business leaders who want to achieve top safety, excellent customer service, and affordable travel for their customers and, at the same time, conduct this in a business-sensible way. Managing a successful airline entails applying strategic management thinking in a challenging industry setting on a daily basis.

A great lesson any newcomer to the airline industry can learn is that in order for an airline to maintain the highest standards of operational safety and quality of service, and ultimately be successful in their strategy, is to invest heavily in the people who will carry out this strategy. Investments in sophisticated equipment and technology alone are not enough to sustain a successful strategy; they have to be coupled with investments in effective human capital development.

Innovative efforts by academics in shedding empirical light on the strategic management of airlines are useful for practitioners. This book by Drs. Flouris and Oswald presents a well-balanced and realistic look at the application of strategic management theory on the airline industry, highlighting the complex character of the industry. The authors' assertion that a poor or misdirected strategy can be disastrous for organizations in a challenging industry such as aviation is something that I find extremely applicable to my role as an industry executive on a daily basis.

Christian Mandl
Chief Executive Officer
SkyEurope Airlines

Preface

This book is an application of strategic management in the aviation industry and the academic field of aviation management. The objective is to cover, effectively and engagingly, what every student of aviation management or of the aviation industry needs to know about crafting and executing business strategies both theoretically and in terms of their practical applications to aviation.

The hallmark of this book is its coverage of strategy related changes in the aviation industry, which is being driven by globalization and the technological revolution. We have taken great care and effort to try and meet the market's need for a comprehensive and multifaceted teaching/learning package that squarely targets what the student of aviation needs to know about crafting and executing business strategies. Consequently, we are also including applied case studies on several airlines and aviation businesses in order to demonstrate how these organizations have addressed strategy formulation and implementation in several different areas, which include corporate strategy, generic strategy, competitive strategy, internal and external environment assessment, mergers, alliances, and safety and security.

Chapter 1 establishes the groundwork for the study of strategic management in the aviation industry. It examines the importance of corporate values and how these values should be reflected in a mission for the organization. The mission is presented as the focal point of the strategic management process, and components of a mission statement are presented as well as sample mission statements. This chapter discusses how, in turbulent, highly competitive environments, the mission might evolve from an analysis of the environmental influences. While this is counter to the traditional strategic management model, a discussion of why this may be a proactive strategic move is presented. Michael Porter's Generic Strategies are introduced with examples from the airline industry, and a short discussion follows as to how Generic Strategies provide a business direction. The chapter concludes with the introduction of objectives, specifically how they are derived from the mission and purpose of objectives in corporate strategy. This discussion introduces competitive strategies as the vehicle for attaining corporate objectives, which will be examined in detail in other chapters.

Chapter 2 further develops the discussion of Porter's Generic Strategies: low-cost leadership, differentiation, and focus or niche strategy. It also studies how combinations of these strategies are often more effective. In addition, this chapter introduces the idea of strategies based on competencies, suggesting that perhaps Porter's model is not as effective in today's competitive environment. The term core competency is introduced, and a discussion follows on how these competencies are a means of focusing on the future and next-generation products and services.

Chapter 3 explains the need for competitive strategies as a means of obtaining corporate objectives and describes how a successful company may not always have a competitive advantage. The idea of competitive capabilities, which allow companies to build a unique position in the marketplace, is introduced. Core competencies build a capability that is not easy for the competition to imitate, leading to a discussion of strategy sustainability and the difference between gaining a competitive advantage and sustaining the advantage. The chapter concludes with a case illustration entitled Institutionalizing Competitive Advantage: Southwest Airlines' Unique Advantage.

Chapter 4 examines how the external environment is a key factor in strategy selection. The emphasis is on monitoring these strategic factors and utilizing all possible sources of information in determining a strategic direction. First, a discussion of the macro external environment is provided. The macro environment includes a detailed discussion of economic, political, environmental, and societal issues. Then a discussion of the micro environment is presented, which centers on an industry assessment and includes information on industry regulations, the competition, the market, the supplier market, and the customer market. The discussion also emphasizes the importance of technology in competitive markets. Michael Porter's Five Forces of Competition model is studied in detail. The idea of identifying niche areas in the market is discussed as a means of competitive advantage and an example is provided from the airline industry. The SWOT (Strengths, Weaknesses, Opportunities, Threats) analysis method is introduced, which will be discussed in the following chapter. This chapter concludes with a case illustration entitled The Denver International Airport: An Environmental Debacle

Chapter 5 completes the second half of the SWOT analysis, the internal environment of the company. The internal environment is further divided into a discussion of the company environment and the individual departments and how strategic decisions must be based on the capabilities of the organization. The analysis emphasizes that a corporation's resources are not only measurable assets, such as bricks and mortar or financial position, but also the skills and competencies of the people within the functional area. A corporation's structure, capabilities, culture, and corporate resources are further highlighted through a study of process management. Michael Porter's Value Chain analysis as a means of assessing the overall process of providing airline services is explained in detail. Further, the idea of benchmarking both within the industry and outside is developed. Emphasis is placed on the importance of financial analysis through ratio comparisons with industry norms as well as closest competitors. Finally, strategic performance is discussed, examining such questions as the following: how well has the company fulfilled its objectives and lived up to the mission, and have they gained a competitive advantage? The chapter concludes with a case illustration entitled JetBlue: Value Added.

Chapter 6 introduces the idea that corporate strategy is based on a combination of a company's orientation toward growth, the assessment of the external environment, and the capabilities of the company. Three general categories of strategy are identified: growth, no growth, and international. The discussion of growth strategies is further divided into types of growth strategies: aggressive, integrated, stabilization, single business, and diversified strategies (related and unrelated). Portfolio analysis is examined as a means of managing various product lines and business units. Examples are provided of each as well as an explanation as to when is the most appropriate time to use each type of strategy. No growth strategies are further divided into liquidation, divestiture, and retrenchment, with explanations and examples. Turnaround is discussed in relation to retrenchment as a means of going from a no-growth position to one of stabilization or growth. The chapter concludes with a discussion on international strategies, including the reasons in favor of expansion and some of the constraints companies face and a case illustration entitled AA's Acquisition of TWA: Timeline of Events.

Chapter 7 examines how the strategic decision-making process, introduced in Chapter 1 and followed throughout the previous chapters, leads to strategy establishment. It suggests that once the internal and external environments are assessed, it is important to review and, if necessary, revise the corporate mission before proceeding to identify strategic alternatives. The importance of SWOT in strategy establishment is examined, as well as the identification of critical issues. The use of contingency plans and scenario analysis is also briefly discussed as a tool of strategy formulation in turbulent environments. The discussion of strategy feasibility once again emphasizes the importance of core competencies to competitive advantage and future sustainability. The chapter concludes with a case illustration entitled Background to Air Transport Regulation and Attempts for Liberalization.

Chapter 8 emphasizes the point that successful strategy does not always equal successful implementation. Strategic buy-in is emphasized as well as the importance of senior management support. Constraints to implementation revolve around the idea of managing the change process. The idea of fitting structure to strategy is discussed and a brief study of organizational structure follows. Further, the idea that implementation follows the basic principles of management, including planning, directing, motivating, and staffing, is explored. The concept of management re-engineering is briefly discussed. The chapter concludes with a discussion of specific implementation problems in the international arena, including commitment of management both at the corporate office and the local office and how local customs and traditions could provide additional implementation constraints.

Chapter 9 describes the final step in the strategic management process: evaluation. The idea of assessing strategic success with regard to both the objectives met as well as the corporate mission is emphasized, which once again returns to the idea of the risk position of the company and the importance of continuous monitoring and the need for scheduled feedback. The link between monitoring and modification is discussed and how modification is often the result of environmental changes both internal and external, which then leads to a discussion of getting back on track and how bad decisions, if not caught soon, could lead to serious problems. The chapter concludes by noting that the strategic management process is a continual process.

Acknowledgements

This manuscript emerged out of our teaching experiences at Auburn University both in aviation and in management. We are indebted to this generous university for its academic vigor and rigor and for the fact that it offers undergraduate and graduate study in aviation management in the College of Business.

Several graduate and undergraduate students, both in aviation and in management and at Auburn University and at Concordia University in Montreal, Canada, have been very helpful in the realization of this book due to their research efforts. From Auburn University we would like to thank Ryan Simpson, Ashley Simpson, Craig Davis, and Tiffany Dunlap, and from Concordia University we would like to thank Kevin Carillo, Herve Riboulet and Serihy Petrenko.

In addition, we would like to offer a special acknowledgement to Gilbert George, a former student in the Global Executive MBA Program in Aviation at Concordia University and management executive for Jet Airways, who allowed us to use his insightful information on Jet Airways as part of this book. Further, we would like to give credit to our colleague, Steve Swidler, who was my co-author in our jointly authored paper on the financial status of the AA-TWA merger.

My friends Adrianus (Dick) and Margaret Groenewege have been of tremendous help for their encouragement and for allowing me to use part of their excellent work, *The Compendium of International Civil Aviation*, on the discussion of aviation liberalization.

Many thanks to Christian Mandl of SkyEurope, who wrote the Foreword to this book despite his very busy schedule.

Special thanks to the editorial team and staff at Ashgate, both in the United Kingdom and in the United States, for their professionalism, encouragement, and support throughout the authoring of this book, especially John Hindley, who believed in the idea of this book, and Guy Loft for seeing it through to its completion.

Lastly, I would like to express my gratitude and appreciation to Joy Leighton. Without her editorial help this book would not have been possible.

Triant Flouris

List of Abbreviations

AA	American Airlines
ACI	Airports Council International
AF	Air France
ALPA	Air Line Pilots Association
AMA	American Management Association
AMR	AMR Corporation (holding company owner of American Airlines)
AOC	Air Operator Certificate (Air Operating Certificate)
AQR	Airline Quality Rating
ATA	American Trans Air
ATC/5	Fifth ICAO Air Transport Conference
ATR72	Avions de Transport Régional, build by Aerospatiale
ATS	Air Traffic Services
A320	Airbus 320
BA	British Airways
BAC1-11	British Aircraft Corporation 1-11 (a.k.a. One-Eleven)
Bn	Billion
B2b	Business to Business
B2c	Business to Customer
B2e	Business to Employee
B727	Boeing 727
B737	Boeing 737
B747	Boeing 747
B757	Boeing 757
B767	Boeing 767
CAD	Canadian Dollars
CASM	Cost Per Available Seat Mile
CDG	Charles de Gaulle airport (France)
CEO	Chief Executive Officer
CRM	Customer Relationship Management
CRS	Computerized Reservation System
DC9	Mc Donnell Douglas DC9
DHL	Dalsey, Hillblom and Lynn, Global Logistics Company
DOT	Department of Transportation (U.S.)
DSS	Decision Support Systems
EDI	Electronic Data Interface (Interchange)

EDIFACT	Electronic Data Interchange For Administration, Commerce and Transport
EC	European Commission
ERP	Enterprise Resource Planning
EU	European Union
EUR	Euro
FAA	Federal Aviation Administration (U.S.)
FASB	Financial Accounting Standards Board
FedEx	Federal Express
FRA	Frankfurt Airport (Germany)
FSC	Full-Service Carrier
GATS	General Agreement on Trade in Services
GDP	Gross Domestic Product
GDS	Global Distribution System
GF-X	Global Freight Exchange
GmbH	Gesellschaft mit beschränkter Haftung (the German equivalent of a limited liability corporation)
GPS	Global Positioning System
GSA	General Service Agent
IATA	International Air Transport Association
IATA AGM	IATA Annual General Meeting
IBM	International Business Machines Corporation
ICAO	International Civil Aviation Organization
IPO	Initial Public Offering
IT	Information Technology
J31	Jetstream J31
J41	Jetstream J41
KLM	Royal Dutch Airlines (Koninklijke Luchtvaart Maatschappij)
LCA	Low-Cost Airline
LCAG	Lufthansa Cargo Charter Agency GmbH
LCC	Low-Cost Carrier
LF	Load Factor
LH	Lufthansa German Airlines
LHR	Heathrow Airport (England)
LSG	LSG Sky Chefs (Lufhtansa Service Holding GmbH)
LUV	Southwest Airlines' symbol at NYSE
L1011	Lockheed 1011
MBA	Master's of Business Administration
MD80	McDonnell Douglas MD80
MIS	Management Information System
MRO	Maintenance Repair Overhaul
NASA	National Aviation and Space Administration (NASA)

NIST	National Institute of Standards and Technology
NYSE	New York Stock Exchange
OSL	Gardermoen airport (Norway)
Q&A	Question and Answer
PSA	Pacific Southwest Airlines
Rt	Round trip
R&D	Research and Development
ROI	Return On Investment
RPK	Revenue Passenger Kilometres
RPM	Revenue Passenger Miles
Sabre	Semi-Automatic Business Research Environment
SARS	Severe Acute Respiratory Syndrome
SAS	Scandinavian Airline Systems
SBU	Strategic Business Unit
SSG	Shared Services Group (Boeing)
STN	Stanstead airport (England)
SWA	Southwest Airlines
SWOT	Strengths Weaknesses Opportunities Threats
TCAA	Transatlantic Common Aviation Area
TRF	Torp airport (Norway)
TWA	Trans World Airlines
U.K.	United Kingdom
ULD	Unit Load Demand
UNCTAD	United Nations Conference on Trade And Development
UPS	United Parcel Service
U.S.	United States of America
USD	United States Dollars
WATS	World Air Transport Summit
WTO	World Trade Organization
XML	Extensible Markup Language

Chapter 1

The Essence of Strategy

> A major problem [with strategy] is partial ignorance; at decision time we do not have the assurance that all the forthcoming attractive activities have been identified and described.
>
> Igor Ansoff[1]

In 1965 when Igor Ansoff said strategy was based on partial ignorance, the field of strategic management was in its infancy. Four decades later, in an era of technological advancements and knowledge management one would expect strategy to be more scientific, more certain. Not so! Yet despite the lack of total information, the art of designing, implementing, and executing corporate strategy is an essential component of strong and effective business management. It is the heart and lungs of corporate competitiveness.

What is Strategic Management?

Strategic management is a planning process used to help an organization determine its mission and general direction or strategic intent. It is a process that helps an organization establish a plan in which to achieve and implement its objectives and which is necessary in order to take a company into the future.

Strategic management provides the path or direction that the organization will follow as it seeks out new markets, attracts or maintains customers, competes with market rivals and conducts the operations of the business. In a sense, strategy is the behavior of an organization. Like any other behavior, strategy is influenced by outside forces as well as internal values, the company's culture and its capabilities. These internal values and capabilities should be the part of the equation that is well known to the organization. These capabilities include such things as financial performance, market performance, safety records and operational performance. However, in an ever-changing industry, like the aviation industry, the outside forces often form the basis for the "partial ignorance," to which Ansoff referred. While some outside forces can possibly be anticipated, such as industry mergers, acquisitions and bankruptcies, oil and gas prices and the economy, nothing could have anticipated the catastrophic

---

[1] Igor H. Ansoff, *Corporate Strategy: An Analytic Approach to Business Policy for Growth and Expansion*, New York: McGraw-Hill, 1965, 152.

incident of September 11, 2001 (also referred to as 9-11) and the subsequent havoc it wreaked on the airline industry specifically as well as throughout the entire worldwide aviation industry and system. This single event created a ripple effect spreading throughout the entire aviation industry. Airlines, airports, air navigation service providers, and general aviation would never be the same after 9-11. New rules and regulations were imposed reaching, for instance, as deep as heightened insurance premiums for flight schools and fixed base operators. System security became a core issue and was addressed in various ways and by various countries and regions, all the way from the local to the international level through world organizations such as ICAO (International Civil Aviation Organization), IATA (International Air Transport Association) and ACI (Airports Council International).

What is the Strategy of an Organization?

The strategy of an organization is based on a series of managerial decisions supported by quantifiable information yet hedged on good judgment. Strategic management attempts to identify the issues that will be important in the future, and based on the capabilities of an organization it provides a roadmap for future behaviors. Today, more than any time in the past, strategy is essential to corporate survival. Why? Because strategy forces managers to think ahead, to be proactive in sculpting the business of the company. Without a strategy, a company is like a piece of driftwood floating through the marketplace with no apparent direction. Companies without strategies generally end up reacting to competitive moves and follow a path of mediocrity, and in a competitive marketplace there is no room for mediocrity. Strategy is also essential to organizational performance; it is the glue that holds all the pieces of the organization together. If all the parts of the organization are not supportive of one another – operations supportive of marketing, finance working with operations, R&D coordinating with operations and information systems coordinating all organizational efforts – the outcome can be disastrous in the marketplace. A well-designed strategy provides the framework for operational rationale.

Yet while strategy is a necessity to good business management, it does not guarantee a problem-free existence. An individual can design the most brilliant strategy ever conceived, but, without the proper implementation, the strategy can be a complete failure. Proper implementation relates back to the organization's internal capabilities. The proper resources – people, money, materials, facilities – must be in place to support and implement the strategy. Corporate success is not always realized immediately, and quick fixes are rare. Often, it takes several years for a good strategy to take hold and for a company to find its optimal position in the marketplace. Sometimes poor market conditions or economic downturns might temporarily delay the expected results of a well-executed strategy, but these should be just this, temporary. These delays should never be excuses for continual mediocre performance. If mediocrity is constant, this may be the sign of poor managerial decisions or hasty decision-making that is not supportive of the strategy. A well-designed strategy should withstand the ups and downs of business cycles without being completely altered. Consider, for instance, Southwest Airlines in the months following 9-11, when the major (legacy) airlines in the U.S. were in cut-back mode, losing money and worried about their very survival. Southwest was operating in a manner that was not much different than any other time in their existence. Their strategy was solid and their implementation well-executed; this is the foundation for marketplace success. Southwest was insulated from the rapid swings of an international network and had a low-cost base of operation coupled with a very unique and effective human resource system.

The Meaning of Competitive Strategies

Having a competitive strategy (as distinct from just any strategy) means that a company is doing something different, deliberately choosing a different set of activities than its competitors.[2] Once again, Southwest Airlines is a good example of a company that has done just this. It originally chose a competitive strategy that was different from its rival firms. Southwest mostly avoided (and still avoids) large, congested airports and offers short-haul, point-to-point service between mid-size cities and secondary airports in large cities. It does this while keeping costs low, which, in turn, helps keep ticket prices lower for consumers. In many cases, because their fares are low and the frequency of the flights are high, Southwest is able to attract people who might otherwise travel by car. Thus, having a competitive strategy means doing something different, offering different activities than competitors. If a company does not have a competitive strategy, then what is the purpose of having any strategy at all? Without a competitive strategy, one company is no different from another.

Southwest's competitive strategy is that it has chosen to be different from full-service airlines. Full-service airlines serve a large number of cities. A large percentage of the customers have connecting flights, and, therefore, full-service airlines have to coordinate schedules and check baggage through to a connecting location. Most of the full-service airlines use a hub-and-spoke method. Under the hub and spoke method, the airline clusters around peak flying hours at hub airports. Southwest, on the other hand, caters primarily to short-haul routes and uses a rolling hub (scheduling) system, where flights leave spaced out throughout the day rather than at a small number of peak times. They also have a standardized fleet of Boeing 737s (B737), which makes maintenance easier and reduces the cost of crew training as pilots, flight attendants and mechanics work on only one type of aircraft. Through its policy of standardization, Southwest has achieved a comparative advantage in

---

[2] Michael E. Porter, "What Is Strategy?," *Harvard Business Review*, 74, no. 6, 1996: 61–78.

the use of the B737, a strategy that is difficult to match. Table 1.1 shows cost per available seat mile for several carriers in the U.S. Fleet standardization is one of the factors that contributes to lower costs for carriers such as Southwest and JetBlue, the latter, which at the time of the publication of the table, had standardized fleets as well. Interestingly, in 2003, JetBlue decided to forgo the benefits of a standardized fleet and ordered a second aircraft type, operating under the assumption that the new aircraft (E-190) would be more efficiently deployed in several markets and the benefits from this more efficient deployment would outweigh the costs of adding a second type in the airline aircraft mix.

Table 1.1 Cost Per Available Seat Mile in 2002[3]

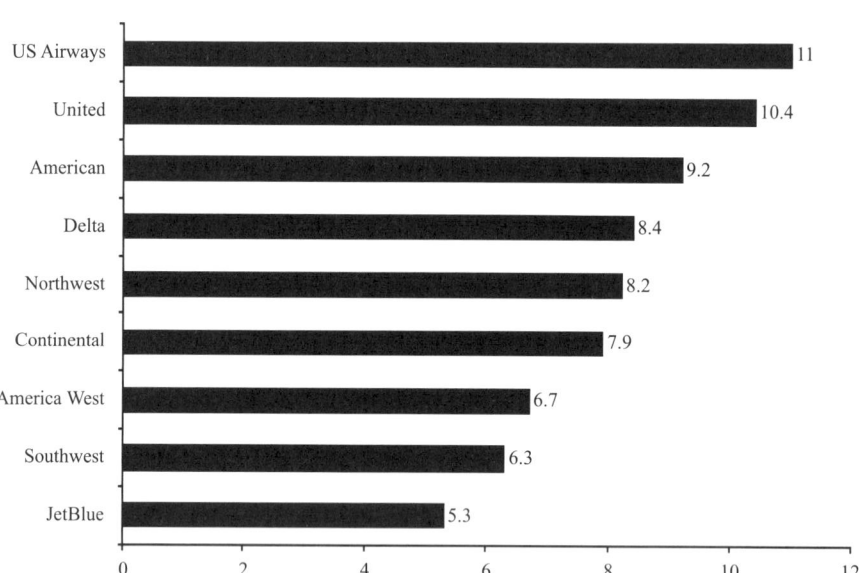

The Difference between Operational Effectiveness and Strategy

It is a given that if a company wants to survive in its business environment, it has to be operationally effective. To be operationally effective one must perform similar activities better than one's competitors. This may mean that a company is more

---

[3] Shawn Tully, "The Airlines' New Deal: It's Not Enough," *Fortune Magazine*, 28 April 2003, 79–81.

efficient or has the ability to keep costs of operation lower than a rival firm's. In fact, operational effectiveness may be the reason that there are differences in the profitability among firms. However, if the activities are similar, but only better, these activities can be easily copied. If they are copied, where does this leave the company? Possibly no better than its rivals. Strategy or competitive strategies, as we have coined it, means that a firm is performing different activities from its competitors or it is performing the same activities in different ways. This is what Southwest and JetBlue did in the United States and Ryanair and EasyJet did in Europe. For a firm to continue to outperform its rivals, it must establish a difference and it must hang on to this difference, preserving it by always thinking ahead. How does a company do this? It must engage in the process of strategic management.[4]

The Strategic Management Process

A corporate executive once said, "Three years ago, when our speaker didn't show up at our quarterly sales meeting, we decided to fill the time by doing some strategic planning. That was the best four hours we ever spent." It would be great if planning was this easy and a company only had to plan once. While it seems like common sense that this would not work, there are managers who never really seem to understand that strategic planning is an on-going process. Strategic management is based on the concept that companies must continually monitor and re-monitor both internal and external influences so that changes can be made to the strategic direction as needed. What happened to the corporate executive's company? It closed two years after he made this statement.[5]

There are six major tasks in the strategic management process: establishing a corporate vision and mission; setting corporate objectives; determining strategic alternatives; selecting a strategy that supports the corporate values, vision and mission; implementing and executing the strategy; and evaluating, monitoring and initiating corrective actions where necessary. While the stages seem clear cut, there is a great deal of interdependence and overlap of the steps. The corporate values, vision, mission and objectives are generally crafted such that each one is reflected in the other. In other words, the values of a company help shape its vision and mission. Yet even these tasks cannot be accomplished without a clear picture of the company's internal capabilities and the environmental forces affecting the company and the industry in which it exists. Without an honest assessment of the company's capabilities, implementation will be compromised. While the strategy process does not require a doctorate, it is dependent on accurate information, well thought-out assumptions, honest evaluations, and sound judgment.

---

[4] Ibid.
[5] Michael Plumb, interview by Sharon Oswald, 15 July 1991.

- *Establish Corporate Values through a Vision and a Mission*: Corporate values are the cornerstone on which the company resides; they are what the company believes in. A strategic vision is an articulation of the company's desired future by organizational leaders. The corporate mission provides a more detailed account of that vision, answering the questions, "Where are we going?", "Who are we?" and "What are we doing?"
- *Set Corporate Objectives*: A company must set specific performance targets within the realm of the corporate vision and mission. In order to be worthwhile, corporate objectives must be measurable and attainable, and they should answer the question, "Where do we want to be in year X?"
- *Determine Strategic Alternatives*: Strategic alternatives come from an analysis of the external and internal environments. Strategic alternatives are the vehicles used to achieve corporate objectives, and they answer the question, "How are we going to get there?"
- *Select Strategy to Support Corporate Mission and Objectives*: The selected strategy should reflect the corporate values as well as achieve the corporate objectives. The selected strategy should be the vehicle that improves the competitive position of the company.
- *Implement and Execute Strategy*: Implementing and executing the strategy means getting all the resources in order and then putting the strategy in place.
- *Evaluate, Monitor and Adjust*: A company must evaluate their performance through constant monitoring and make adjustments when new developments warrant corrective actions. Sometimes adjustments require alterations in the mission, objectives, strategic alternatives, and selected strategies. The importance of environmental scanning cannot be emphasized enough.

This strategic management process establishes the framework for the upcoming chapters.

The next section will focus on corporate values, vision, and mission and their importance to the strategy process. While all three appear to overlap, individually they each help to form the strategic purpose of the organization and the foundation on which it operates.

Corporate Values and Social Responsibility

The topic of corporate values, particularly with regard to ethical behavior and social responsibility, has received a great deal of attention in the aftermath of the Enron and WorldCom debacles. Corporate values are considered by some to be the core element in the drive to become competitively dominant. These values are the uncompromising guiding principles of a company and are used to motivate employees and to give employees something in which to believe. In effect, these values shape organizational

> **Values:**
> The Principles of the Organization
>
> **Vision:**
> The Future of the Organization
>
> **Mission:**
> The Distinctive Characteristics of the Organization

attitudes and employee behaviors. Yet sometimes corporate values are little more than a few words, often clichés. For example, annual reports throughout the world use such statements as "integrity," "customer service" and "excellence." These are represented as the values of the company, the essence of their existence.

Corporate values are a set of principles that guide and define how a company should treat its employees, shareholders and customers. By definition, corporate values should do the following:

1. Remain intact, regardless of changing business cycles or economic swings.
2. Be relevant and applicable to every person in the organization.
3. Be applied consistently, regardless of issues.
4. Remain constant, regardless of staff or management changes.[6]

For corporate values to play a meaningful role in organizational effectiveness and for these values to form the foundation for creating and maintaining the organization, corporate values must be first-order. In other words, corporate values should be derived from a fundamental philosophy about what actually constitutes the good of the organization. The values are management's attempt to define the organizational good both in the context of organizational life and in all of the actions and endeavors in which the organization engages. First-order values should be the very reason the organization exists; in essence, they are a constitutional framework for corporate governance.[7]

Many corporations take these first-order values one step further by situating the good of the organization in a societal context. This is known as social responsibility. Social responsibility has become an important value for many companies trying to focus on moral and societal issues and events. Social responsibility revolves around the idea of giving something back to society, whether it is in terms of monetary

---

[6] Sandy French, "CEO Values Replace Corporate Values," *Canadian HR Reporter*, 8 April 2002, 4.
[7] Edward J. Giblin and Linda E. Amuso, "Putting Meaning into Corporate Values," *Business Forum*, 22, no. 1, 1997: 14–19.

donations, sponsorship of events, or ethical words by which to live. Much of the current interest in social responsibility supports the belief that an ethical overtone is in the best interest of the company. It is a type of Hippocratic oath for a business, which tells the customer exactly what to expect from the company.

Even more important today is the issue of ethical behavior as part of the company's core value system. Company stakeholders want to see a written commitment to ethics in business dealings. A 2002 study by the American Management Association (AMA) indicated that 76 percent of the 175 executive respondents listed ethics and integrity among their core corporate values. In addition, 86 percent reported that they had written corporate values that were disseminated to employees through handbooks, brochures, wall posters, or website. However, despite the emphasis on corporate values, the same study revealed that 32 percent of these companies did not necessarily follow their corporate values and that their written word was often quite different from their actions.[8] Many people might find this last statement very disturbing. What is the purpose of having corporate values if they are not followed? The corporate values, vision, and mission should provide the foundation on which an organization is built and moves into the future.

Designing a Corporate Vision

A strategic vision is an articulation of the company's desired future by organizational leaders; it is truly an integral part of the strategic management process. One of the first important steps in the strategy process is establishing a corporate vision. Top managers are responsible for crafting the corporate vision. It should answer such questions as the following:

- Where should the company be in the future?
- What is the direction in which the company is headed?
- What place should the company occupy in the marketplace?

Establishing a vision requires management to take a close look at the company, its capabilities, its product, and its competitiveness. It forces managers to focus outside of the organization on the market and the industry and make some tough decisions as to where the company can best succeed. The vision reflects the aspirations and ambitions of the top managers with respect to the company, and a strong vision means that it becomes the guiding perspective and driving force of the organization. Put succinctly, a strategic vision should provide a sense of commitment and cohesiveness for the organization.

---

[8] American Management Association (AMA), *AMA 2002 Corporate Values Survey*, New York: AMA, 2002, see <http://www. amanet. org/research/pdfs/2002_corp_value. pdf>.

Westley and Mintzberg suggested that to establish a true vision within a company three very distinct steps must be followed:[9]

- Step one: Envision an image of where the company should be in the future.
- Step two: Articulate this vision to everyone who is affected by the vision, employees, customers and any followers such as stockholders.
- Step three: Empower these individuals, employees, customers and followers to enact, or "live," the vision.

One of the biggest problems found in business is that all too often the vision is nothing more than a statement that top management is aware of but to which the employees, customers and even the stockholders may never have been exposed. As a result, no one knows the one statement that is supposed to "rule" the future events of the company!

A study by the National Institute of Standards and Technology confirmed just this. Prepared by Harris and Associates, the study of 300 executives from large U.S. industrial and service companies revealed that while 82 percent of the individuals interviewed said that they had a definite vision of what they felt their organization should be in the future, and 79 percent of them thought that a long-term vision was necessary for the survival of their companies, only 38 percent felt that their visions were broadly shared within the organization.[10] If a vision is never articulated, it cannot be very clear to anyone outside of top management offices. If it is not very clear to anyone outside of top management, how are the employees supposed to believe in what the organization is all about? Logically, if the vision is not shared and, therefore, people are not aware of it, then how can a company say that a vision exists? Westley and Mintzberg said that if vision is to exist it must be the result of an interaction between the leaders and followers and the vision should empower everyone involved.[11]

The vision forces managers to think into the future. If a manager does not know where he or she wants to be in the future and does not know the direction to steer the business, how can he or she run the business? The vision provides the future path for the business. Therefore, the strategic planning efforts should support the vision and also inform the organization's vision. What does this mean? Through planning efforts, managers become aware of the environment around them. As the environment changes (which will be discussed in Chapter 4), managers will begin to realize the effectiveness of their vision.

---

[9] F. Westley and H. Mintzberg, "Visionary Leadership and Strategic Management," *Strategic Management Journal*, 10, 1989: 17–32.

[10] National Institute of Standards and Technology (NIST), 818407 (2003), see <http://www.ddworld.com/pdf/LADFSO2R.PDF+Harris+and+Associates+study+on+corporate+vision&hl=en>.

[11] F. Westley and H. Mintzberg, 17–32.

It is evident that a corporate vision cannot be designed in a vacuum. Unless managers thoroughly study the marketplace, they may not be able to see the future clearly. A thorough examination of the market tells a manager what the competition is doing, what the customers want, what new innovations are on the horizon and what the technological requirements are now and will be in the future. Setting a future direction must be based on an honest assessment of the market. Without the knowledge of the marketplace, managers do not know whether the company can sustain itself in the future or if the business needs to change in order to better meet future demands and needs.

Is a Strategic Vision Really Important?

Imagine as a pilot getting into an airplane without a flight plan; the pilot knows what is needed to fly to a specific destination, but he or she does not know the specific route to take to arrive at the goal. A strategic vision is like a flight plan. While corporate values provide the moral and ethical fiber for a company and provide the incentive which serves to motivate its employees, a firm's strategic vision has much more direction-setting value. Consequently, a firm's vision has more of a direct link to its status in the marketplace.

Managers are required to keep their finger on the pulse of the future. Every day new technologies are developed and these new innovations may have an effect on the way a company conducts business in the future. How might the company be affected? It might see changes in the needs or expectations of its customers or realize that the changing marketplace could make current business practices obsolete. It might see a path toward competitive dominance or discover that the future holds new and better ways to provide services, such as reducing overhead or labor costs.

To be successful in business today, managers must be forward thinking. A vision statement forces managers to think about the future. When top management fails to look to the future, they are committing their company to a dated strategy, which could lead to the demise of the company or, at the very least, place it in a reactive market position.

Sample Vision Statements[12]

AirNet Express (Overnight Freight)
"AirNet is committed to focusing its resources on providing value-added time-critical air shipment and aviation services to a diverse set of customers in the most service-intensive and cost effective manner possible."

---

[12] See <http://www.raytheon.com/about/gva.htm>; http://www.amrcorp.com/corpinfo.htm; http://www.heliwing.com/; and <http://www.airnet.com/Company/companyFrame.htm>.

> **AMR (American Airlines)**
> "To be the world's leading airline by focusing on industry leadership in the areas of safety, service, network, product, technology, and culture."
>
> **Heliwing Helicopters Ltd.**
> "To continue to develop strategies and incorporate new technology to ensure our company remains progressive while maintaining our record of safety and reliability. Our operations shall continue to be conducted with integrity and dedication to service."
>
> **Raytheon (Owners of Beechcraft)**
> "Be the most admired defense and aerospace systems supplier through world-class people and technology."

Designing a Mission Statement

The mission is a broadly defined statement of purpose. Mission statements are called many different things such as creeds, purposes, or statements of corporate philosophy. No matter what term is used, a mission statement defines the business in terms of scope and purpose. While the mission is broad, it is, in some ways, also specific in nature. A mission must be broad enough to allow for innovation and expansion, to allow for natural organizational growth, yet narrow enough to establish some direction for the company. For example, the corporate mission generally answers the questions, "Who are we?", "What do we do?" and "How do we differ from the competition?" A mission statement is the means by which an organization distinguishes itself from all others. The mission is like a mirror: it reflects how the company sees itself, yet it also sets the parameters for future business decisions, and this is why it needs to be somewhat broad. (Several sample general, functional area and project specific mission statements are provided at the end of Chapter 1.)

The mission of a company should not change unless there is a dramatic change in business operations. Again, it should be limited enough to set the parameters but broad enough not to require constant change. What are the most common elements of a mission statement? The results of a 1994 *Business Week* study of corporations showed that companies with mission statements included concern for public image (73 percent); concern of quality (73 percent); commitment to survival, growth and profitability (70 percent); identity of customer and markets (60 percent); identity of products and services (60 percent); statement of company philosophy (43 percent); and differentiation from the competition (33 percent).[13]

---

[13] Charles Rarick and John Vitton, "Mission Statements Make Cents," *Journal of Business Strategy*, 16, 1995: 11–13.

> **What Questions Should be Answered by the Mission Statement?**
>
> - Who is the customer?
> - What is the product?
> - What market is being served?
> - What makes this company different from the competition?
> - What are the beliefs, philosophies, and values of the company?
> - What is the level of commitment to survival, growth and profitability?
>
> **Additional items to include:**
>
> - Statements of technological advancement.
> - Concern for the corporate image.
> - Inspiring qualities of the company.
> - Statements of social responsibility.
> - Concern for quality.

The Reason Mission Statements are Important

Mission statements have been found to increase shareholder value. A *Business Week* study of 1,000 corporations indicated that having a corporate mission statement had a favorable impact on corporate profitability, boosting shareholder equity. Corporations with mission statements had an average return on stockholder value of 16.1 percent, while those without mission statements reported only a 7.9 percent average return.[14] Why would a mission statement have this kind of effect? If the mission statement provides the framework for corporate decisions and sets the parameters for corporate moves, it gives focus to the company, something which is essential if it is to be successful competitively.

Mission statements can also be developed to assure projects meet the expectations of the constituents, as was the case with the Oakland International Airport expansion (see end of chapter). It is also not unusual for functional areas to have mission statements. The mission statements of functional areas should reflect the overall company mission. Continental Airlines' "Fly to Win" mission provides the overarching structure for its functional area mission statements (see end of chapter).

Mission Statements and the Environment

In a highly turbulent, ever-changing environment, a company might find it more

---

[14] Ibid.

advantageous to establish its mission after a detailed environmental analysis. In a situation like this, the company might be one that has not been competitive and is looking for an entirely new direction for the organization. Alternatively, it may be a new entrant into the market looking for the one area where it might establish a new way of doing something, a next-generation approach to the industry. In a situation like this, the mission becomes the result of the environmental and industry analysis (discussed in Chapter 4).

Setting Organizational Objectives

The next step in the strategy-making process is to set organizational objectives. The objectives translate the values, vision, and mission statement into specific performance targets. In effect, they solidify the plans of the organization to achieve specific targets. They are a kind of action plan for the company. If it is not attached to specific targets, the mission statement cannot begin to increase shareholder value. These performance targets are then used as milestones of management achievement. They give direction to management in striving to be competitive and in sustaining a competitive posture.

For these objectives to be worthwhile, they must be direct, quantitative or measurable and attainable. Further, they must have a timeline attached to them for achievement. In other words, the objectives spell out how much of something will occur in a specified period of time. Measurement is essential. A company cannot realize its objectives unless they are measurable. For example, if the objectives are stated to "maximize profits" or "increase sales" what does this really mean? Nothing. Instead, the company should use specific, measurable objectives like "increase profitability by 5 percent" or "increase sales by 10 percent."

How aggressive should a business be in setting its objectives? Objectives should be a means of stretching the organization to achieve its full potential. Aggressive, forward moving objectives should not be feared and should push managers to the point of being challenged. Some believe that it is important to set objectives almost to the point of impossibility, but one must be realistic. If one is too aggressive in defining the company objectives, then management will continually face failure, and failure leads to a less motivated management team. On the other hand, one does not want to be too lenient in defining objectives. At the least, minimum incremental improvement is needed, but, depending on the industry and the competitive environment, incremental improvement may not be enough to sustain an organization in the competitive arena.

The appropriate targets for strategic objective are dependent upon a company's internal capabilities as well as the environment in which it exists. It is also dependent on what level of performance will satisfy the customers and shareholders. Most importantly, though, the objectives must be an extension of the mission statement.

14   *Designing and Executing Strategy in Aviation Management*

Long-Range and Short-Range Objectives

When objectives are being established, one should think both in terms of long-range objectives and short-range objectives. In the dynamic environment in which organizations presently exist, long-range generally means five years. Long-range objectives force one to think in terms of the position and future performance of the organization. Whether or not a person is setting long or short-term objectives, he or she must always be aware of what is going on in the world around them. If one does not have long-term objectives, focusing only on the short-term, the goal of competitive sustainability is neglected. Competitive sustainability can only be achieved by always looking to next-generation products or services. By establishing long-term objectives, one is essentially setting long-range performance targets for their managers.

The short-term objectives serve as a means of forcing management to take immediate action toward achieving long-term objectives. They can be the milestones or stepping stones toward achieving the desired state.

Objectives Should be Pervasive Throughout the Organization

It is essential to establish organizational objectives in order to provide direction for an organization. However, objective setting should not stop at the top. In fact, objectives should be set at all levels throughout the organization. The concept of top-down objective setting means that the direction is established at the top, but the divisions, separate businesses, departments and functional areas establish their own set of objectives. The objectives that are set at other levels in the organization are there to support the corporate objectives. For example, if a company says that they want to increase profitability by 5 percent, then each of these areas should have an objective that will help the company achieve this level. Like the organizational objectives, the objectives set at other levels of the organization must be measurable and attainable. These objectives must stretch the organizational units to achieve the highest levels of performance. By setting the organizational objectives at the top and then allowing organizational units to establish complementary objectives, it determines both the strategic and financial direction of the organization.

Strategic Versus Financial Objectives

When one thinks of objectives, generally the first thing that comes to mind is financial objectives, for instance "we will increase profitability by 5 percent in two years." Financial objectives are essential to any organization: one cannot look to the future without financial objectives like earnings, revenues, dividends and stock prices. If one does not have acceptable financial performance, then there is little chance that any kind of strategic objective will be achieved.

Strategic objectives are those objectives that are directed toward issues like market position, product quality, customer service, geographic coverage and cost reductions. They are essential to attaining and sustaining the company's competitive position in the industry. The strategic objectives help define the strategic intent of the organization, which differs from company to company. If the intent is to be a global competitor, then the strategic objectives should reflect this; if the intent is to gain market share in the domestic market, again, the strategic objectives must state this. Like financial objectives, strategic objectives are essential to the strategic management process. They set the specific performance targets both financially and strategically.

Financial vs. Strategic Objectives for the Aviation Industry

Financial Objectives	*Strategic Objectives*
• US Airways (2003): Successfully emerge from Chapter 11 bankruptcy protection.	• US Airways (2003): Reduce average operating costs to restore competitive situation.
• Air Canada (2004): Bring operational costs under control.	• Air Canada (2004): Establish trust in the relationship of management and employee unions.

Carving Out a Strategy

Now that corporate values, vision, mission statement and the objectives or target for achieving the mission have been established, the next step is strategy development. Strategies are the vehicles for achieving the objectives. They describe how organizational goals will be achieved – how financial and performance targets will be met, how the competition will be outsmarted and how to become a prominent force within the industry.

Professor Michael Porter suggested that a company's competitive strategy consists of the business approaches and initiatives it engages in to attract new customers, withstand the competitive environment and strengthen its competitive market position. While there are countless competitive strategies that companies can embark upon, Porter noted that company strategies can be boiled down to fit into three classifications, or a combination of the three. Known as his Generic Strategies, these strategies are low-cost leadership, niche, or differentiation. Chapter 2 examines Porter's Generic Strategies and why, in the changing environment today, they may no longer be enough.[15]

---

[15] Michael E. Porter, *Competitive Strategy Techniques for Analyzing Industries and Competitors*, New York: Free Press, 1980, 34–46.

Strategy is an On-going Process

Once corporate values, vision, mission, objectives and strategies have been established, the job is still not complete. In fact, the strategic management process is never final; it is an on-going process. If the environment remained stagnant and nothing ever changed, then the same objectives and strategies would suffice forever. But people do not live in a fish bowl! The world is fluid and the aviation world in particular is a very competitive, ever-changing environment, characterized by a history of strict regulation and great sensitivity to the business cycle. A normal part of the strategic management process is to review environmental changes, evaluate the performance of the strategies that have been put into place, and make adjustments as needed. It is up to the managers responsible for implementing these strategies to monitor the situation, keep abreast of the problems as well as the opportunities and be aware of environmental threats. Their duty is to initiate the appropriate corrective actions, whether it is a simple change or a totally different strategic approach. In the following chapters, the steps involved in the strategic management process will be further examined, beginning by looking at how to craft a company's strategies.

Sample Mission Statements[16]

General Mission Statements

Virgin Atlantic: To grow a profitable airline, that people love to fly and where people love to work.

Heliwing Helicopters Ltd.: To provide a professional helicopter service of the highest quality.

TourJet: As an aircraft charter broker, TourJet will access a Worldwide network of aircraft types and operators; ensuring that the best possible aircraft is matched to the mission requirement of Tour Jet's clients.

Functional Area Mission Statements

Continental Airlines:
Fly to Win (Market Plan)
Achieve above-average profits in a changed industry environment. Grow the airline where it can make money and keep improving the business/leisure mix. Maximize

---
[16] See <http://www.virgin-atlantic/our_story_student_packview.do>; <http://www. heliwing. com/>; <http://www.tourjet.net/mission.html>; <http://www.flyoakland. com/tex/terminal_expansion_mission.shtml>; and <http://www.continental.com/com/company/investor/docs/continental_ar_2002.pdf>.

distribution channels while reducing distribution costs and eliminating non-value-added costs.

Fund the Future (Financial Plan)

Manage our assets to maximize stockholder value and build for the future. Reduce costs with technology. Generate strong cash flow and improve financial flexibility by increasing our cash balance.

Make Reliability a Reality (Product Plan)

Deliver an industry-leading product we are proud to sell. Rank among the top of the industry in the key DOT measurements: on-time arrivals, baggage handling, complaints and involuntary denied boardings. Keep improving our product.

Working Together (People Plan)

Help well-trained employees build careers they enjoy every day. Treat each other with dignity and respect. Focus on safety, make employee programs easy to use and keep improving communication. Pay compensation that is fair to employees and fair to the company."

Project Specific Mission Statement

Oakland International Airport, Terminal Expansion: The Port intends to reconstruct the passenger terminal facilities at Oakland International Airport to serve as a major gateway to Oakland and the East Bay and North Bay Regions, with a cost conscious approach, to provide convenient, cost effective facilities with the realization that the theme of the terminal complex should reflect the diversity of the communities served. The complex will be designed to be aesthetically attractive and should present the user with a series of interesting and unique experiences. The use of art in the construction of the complex will be a priority. This terminal complex shall be functional, constructable [sic] and maintainable for all users and shall be appropriate to handle the increasing passenger traffic through the initial years of the twenty-first century. The Airport acknowledges itself as a "best value" airport and as such, these facilities must be built in compliance with the established budget using an open process where information is shared by all. During construction, existing Airport operations must be maintained with only the most necessary service disruptions. To the greatest extent economically possible, commercial opportunities for local businesses will be maximized, consistent with Port Policies throughout all aspects of the Program. Adherence to the above will create a safe, cost effective and secure terminal complex that provides the maximum in customer service to the users of the Airport.

Chapter 2

Strategic Positioning and Sustaining a Market Presence

> When price becomes the most important issue, when you're selling a pure commodity, the company that wins is the lowest-cost producer.
>
> Scott Kirby[1]

> Competitive strategy is about being different.
>
> Michael E. Porter[2]

The key to winning in the competitive arena is to have a sustainable competitive advantage. It is not enough to have a good idea; the idea must be one that can withstand environmental and competitive pressures and sustain technological advancements. A company that has the edge over its competition has to have a defendable position that cannot be easily imitated.

This chapter focuses on competitive strategy and how a company can achieve and sustain a competitive advantage. We will begin by examining what Michael Porter, one of the most influential thinkers in the field of strategic management, calls the Generic Strategies. We will further discuss how these Generic Strategies might not be as true in the global, dynamic environment of today and, in particular, in the international aviation arena.

The Generic Strategies

As discussed in Chapter 1, Michael Porter suggested that a company's overall competitive strategy consists of the business approaches and initiatives it engages in to attract new customers, withstand the competitive environment and strengthen its competitive market position.[3] While there are as many competitive strategies as there are competitors, Porter said that corporate strategies can be boiled down to fit into three classifications, or a combination of the three. Porter's Generic Strategies

---

[1] Scott Kirby quoted in Bruce Schoenfeld, "We're Fed Up!," *Cigar Aficionado*, July/August 2002, 58.
[2] Michael E. Porter, "What is Strategy?," *Harvard Business Review*, 74, no. 6, 1996: 64.
[3] Ibid.

follow three distinct approaches: low-cost leadership, niche, or differentiation. These three approaches strive to carve out a different market position in an attempt to gain that portion of the market. While two of these market positions apply to a broad spectrum of customers (low-cost leadership and differentiation), the niche market can be as successful by exploiting a small corner of the buyer market.

Low-Cost Leadership

This approach is based on being the overall low-cost provider in the industry.
Example airline: Ryanair

Niche

The approach is based on concentrating on a narrow buyer segment and out-competing rivals by offering a customized product or service that meets the requirements of the customers better than other competitors.
Example airlines: Aloha Air (geographic niche), Concorde (customer niche) and Chalk (product line niche)

Differentiation

This approach seeks to provide a difference to the product in the industry, a difference for which the public is willing to pay extra.
Example airline: Singapore Airlines

When America West's Scott Kirby called the airline industry a "commodity industry," he suggested that the general public did not know the difference between airlines such as United, American, Delta, Southwest, and JetBlue. His argument was that the public is driven by price when making purchasing decisions as to which airline to fly. However, price alone does not always explain consumer choices in air transportation. Airline travel can be segmented using several dimensions. One important dimension is that of leisure passengers and business passengers. While price may rank high in the minds of leisure passengers where elasticity of demand is high, when it comes to business travel demand tends to be quite inelastic and depends mostly on amenities of convenience, for example, schedule rather than price. Business passengers tend to be sensitive to prices during periods of economic downturns when companies normally streamline spending. However, they are also concerned with things other than price in situations where distance comes into play. Whereas in-flight amenities may not be very important to a business passenger in a one-hour flight, they may become of considerable importance in a long-haul flight, as such their propensity to pay higher prices increases.

Furthermore, airline transportation can be segmented according to the accessibility of information by a passenger. Passengers may pay higher fares than they would probably like to pay for a specific flight segment because they use a medium to make their reservation that does not have access to the lowest possible price. For example, some passengers may choose a travel agent, direct bookings over the phone rather than the internet, or, in some cases, sites on the internet that do not have the best possible fares available. In the case of price sensitive consumers, this may lead to customers finding an alternative means of transportation.

In other cases, consumers may choose one airline over another airline due to preconceived perceptions of in-flight service, ground services, or safety. Others might select a particular airline for cultural or national reasons (flying a flag carrier over a rival airline from another country or even an airline associated with a specific geographic region) or because they are members of a specific frequent flier program or alliance. Understanding consumer behaviors is important in establishing a competitive posture.

Low-Cost Leadership Position

When buyers are price sensitive, the airline that takes a low-cost leadership posture will have a very strong competitive position in the market. The goal of a low-cost leader is to contain the costs to the lowest relative to industry rivals and, in essence, to create a sustainable cost advantage over the competition. The key to this strategy is that cost is not equal to price. When we refer to cost, we refer to the amount of money or resources it takes to produce the product or provide the service. A low-cost leader will make sure that it offers a product with features that are the bare essential requirements for industry consumers, because without the bare essentials no one is going to purchase the product or service. In order for the cost advantage to be effective and sustainable, the company has to gain the cost advantage in a manner that is very difficult for rival firms to copy. This is the important point to realize: the cost advantage is based totally on the relative ease or difficulty it takes to imitate the product. This is the key to sustainability and the competitive edge.

The generic business model that low-cost carriers (LCCs) have followed is that of low-cost leadership. The main idea has been that lower production costs can be translated to the consumers as lower prices. This has led to price sensitive consumers switching from legacy carriers to low-cost carriers, specifically in situations where the consumer finds the schedule of the low-cost carrier convenient. In addition, low-cost carriers have been able to stimulate traffic in city pairs that consumers would not otherwise fly to because of their offer of lower fares.

There are two options that a low-cost leader has for achieving a distinctive level of profitability. A low-cost leader can make money by providing the service at a price lower than the competition and, therefore, gaining a larger share of the marketplace. In this option the low-cost leader is undercutting the competition. If the buyer is truly cost-sensitive, then the buyer will likely choose the product or service offered by the low-cost leader. Obviously, this works best if the product or service is more of a commodity.

The other option is for the low-cost leader to charge the same price as other competitors. Why is this a good option? In this case, the low-cost leader reaps the benefits of a greater profit margin. How? The low-cost leader by definition has kept the cost of doing business lower than other rival firms. Because the costs are lower, the difference in profit margins between the low-cost leader and other competitors can be significant. This is a very good strategy in situations where low-cost leaders have penetrated a specific market; however, in the case of the airline industry, it does not work as well when the low-cost leader brand is unknown as consumers need incentives to choose the low-cost leader over its competitor, which in most cases is an incumbent airline.

What are the Competitive Strengths of a Low-Cost Leadership Strategy?

- **Rival Competitors**: Better positioned to compete offensively on basis of price.

- **Buyers**: Better protected from negotiating power of large customers.

- **Suppliers**: More insulated than competitors from powerful suppliers.

- **Potential Entrants**: Low-cost provider's pricing power is a significant entry barrier.

- **Substitutes**: Better positioned to use low price as a defense against substitutes.

In the airline industry it is not enough to be a low-cost leader; an airline must also provide the customer with a workable schedule. For instance, consider the conditions in the market from London, U.K., to Oslo, Norway. Non-stop scheduled service from London's Heathrow (LHR) Airport to Oslo's main airport is conducted by both Scandinavian Airline Systems (SAS) and British Airways (BA). In February 2003, the lowest published fare for this market was $145 round trip (rt) and $156 rt respectively. SAS and BA provide service four and five times a day respectively. Low-cost operator Ryanair has tried to enter this market with considerably lower fares ($62.42rt), but its schedule, in this particular example, is considerably underdeveloped. Specifically, to get the $62 fare, the service is actually from London's Stanstead (STN) airport, a Heathrow reliever some 48.5 miles away, nearly an hour by taxi or bus at a fare of approximately $21.52 each way. The twice-a-day schedule does not land the passenger at Oslo Gardermoen (OSL) but actually in Sandefjord, Norway, at what Ryanair calls "Oslo (Torp)." Torp (TRF) is an additional one hour by train from the off-airport train station to Oslo or a cab ride of $14.58 each way. To deconstruct the customer's perceived savings further, the many transatlantic travelers who connect through London Heathrow will not likely be able to use Ryanair schedules as most international arrivals are mid to late morning (adding on the time it takes to get through immigration, baggage, and transfer). As seen in Table 2.1, the departure and arrival times do not work well for such international travel.

Table 2.1 Comparative Schedules and Prices between Ryanair, SAS and British Airways[4]

Originating Flights:

	SAS		British Airways	
London – Oslo	Fare: $145rt		Fare: $156rt	
	Dep.	Arr.	Dep.	Arr.
LHR – OSL	7: 20a	10: 25a	7: 20a	10: 30a
	10: 30a	1: 35p	1: 20p	4: 25p
	12: 50p	3: 55p	4: 15p	7: 20p
	5: 35p	8: 40p	7: 35p	10: 40p
	Ryanair			
	Fare: $62.42rt			
	Dep.	Arr.		
STN – TRF	7: 20a	10: 15a		
	5: 55p	8: 50p		

Return Flights:

	SAS		British Airways	
	Dep.	Arr.	Dep.	Arr.
OSL – LHR	7: 55a	9: 20a	7: 45a	9: 15a
	10: 25a	1: 15p	11: 50a	1: 15p
	3: 00p	4: 25p	5: 25p	6: 55p
	5: 05p	6: 30p	8: 10p	9: 30p
	7: 55p	9: 20p		
	Ryanair			
	Dep.	Arr.		
TRF – STN	10: 40a	11: 35a		
	9: 15p	10: 10		

Note: All fares exclude taxes and have been converted to USD at the exchange rate of 2-5-03.

---

[4] See <http//: www.travelocity.com>; <http: //www.ryanair.com>; <http: //www.torp.no>; and <http//: www.baa.co.uk/main/airports/stanstead>.

24 Designing and Executing Strategy in Aviation Management

When Should a Low-cost Leadership Strategy be Used?

As previously suggested, the low-cost leadership strategy works the best when the buyers in the industry are price sensitive. If buyers base their decisions on who among the rival firms offers the lowest price, then the low-cost leadership position seems to be the obvious choice.

When a Low-Cost Strategy Works Best

- Price competition among rivals is a dominant competitive force.
- The industry's product is a commodity-type item readily available.
- There are few ways to achieve product differentiation that have value to buyers.
- Most buyers have similar needs/requirements.
- Buyers incur low switching cost changing sellers.
- Buyers are large and have significant bargaining power.

If all the industry buyers use the product in the same way, the product is most likely standardized across all industry players. When the product is standardized, issues of quality or even special features do not become important; the price of the product is what is important. Consequently, the low-cost leadership position is attractive because it is one way the firm can attain dominance over rivals in the industry. Likewise, with a standardized product there are few if any ways an industry player can differentiate itself from the other players. Whatever method of differentiation is used, it is probably not of interest to the buyer and, therefore, would not be a smart competitive move. Again, the only real differentiating factor would be the price of the product.

When there are many competitors in the industry and the rivalry is particularly strong, a low-cost leadership position would again be worthwhile when it comes to the prices of the product. In this case the company could undercut the rival firms and still be more profitable because it would be gaining a larger portion of the market.

Another industry situation where low-cost leadership is a good competitive strategy is when the buyers in the industry carry a lot of weight and have the ability to negotiate prices down. The low-cost leader can best satisfy this buyer requirement and still remain profitable.

What do Managers Have to Do to Achieve Low-Cost Leadership?

In order to obtain a low-cost leadership position, management must focus attention on costs at every functional level of the organization. Low-cost leaders typically do the following:

- Optimize the operating efficiency of facilities.
- Pursue cost reductions through tight procedural controls and avoidance or elimination of marginal customer accounts.
- Minimize costs in areas like research and development.

In order to achieve low-cost leadership, managers have to scrutinize each cost-creating activity, which are known as cost drivers. Managers need to be intimate with these cost drivers and know everything about them in an attempt to continue to manage the cost of these drivers year in and year out. At the same time, it is important that managers not become complacent. Gaining the low-cost leadership position might take a fundamental re-engineering of how activities are performed and coordinated. Sometimes it means cutting out steps in the manufacturing process, using next-generation products or robotics to help provide the service, or producing the product at a fractional cost of one's competitors. Sometimes it just means being entrepreneurially creative. For example, Ryanair has gotten rid of window shades in its aircraft as a way to eliminate maintenance costs associated with broken or non-functional shades and to reduce fuel costs by eliminating this additional weight.

The virtual corporation is a recent example of the direction a business can take if it is committed to low-cost leadership. Virtual corporations eliminate many of the key cost components of more traditional competitors. A lot of start-up airlines use this strategy through outsourcing most functions and choosing to lease aircraft rather than to buy them. However, this strategy can be a double-edged sword if it is not coupled with close control. The company that is providing outsourcing must perform the function at a significant cost reduction and in a competent manner, without compromising the airline's core offering. Even a small business with few facilities and employees can establish a low-cost leadership position.

The leadership position comes from two major factors:

- Cost consciousness, where cost is always a prominent consideration in decisions for items like travel, equipment and supplies.
- Adaptability to market conditions, where key cost components (e.g. inventory) can be adjusted rapidly, depending upon demand, supply for the products, or services of the business.

Differentiation Strategies

With a differentiation strategy, the company competes in the marketplace by providing a product or service that is unique in the industry. The uniqueness can be in features offered, different appearances, or anything that adds customer value, or, at least, perceived customer value. A differentiation strategy does not have to add value at all; it just has to be something that the customer perceives to be better or worth paying

> **What are the Problems with a Low-Cost Leadership Position?**
>
> - Technical breakthroughs can negate the low-cost leadership position, because they make cost reductions available for other players in the industry, negating a low-cost provider's efficiency advantages.
>
> - If the leadership position is not grounded in a sustainable advantage, rivals can find it relatively easy or even inexpensive to imitate the costing methods used by the low-cost leader.
>
> - Sometimes the low-cost provider becomes so intent on keeping costs low that it is blind to the changes in the buyer's requirement for added quality or features. The low-cost provider fails to see new developments in related or substitute products and fails to notice that buyer sensitivity to costs no longer exists.

for. The main objective of a differentiated firm is to provide a product or service with some distinguishable feature for which the customer is willing to pay a premium. Thus, if what they are providing is truly considered to be a valued contribution, the customer will pay the extra money. Why is this a good strategy? The premium price paid by the customer is a profit incentive to the company providing the product or service. JetBlue Airways offers a good example of differentiation strategy as applied by a low-cost airline carrier. For example, JetBlue offers DIRECTV in its flights, which is a unique feature not offered by any other carrier in the United States at the time of authoring this book. Passengers have the ability to watch live TV in their airline seats, as they all have a small TV screen in front of them, and the ability to control programming for their individual screens. Even though JetBlue originally planned to charge customers for this feature through a usage fee, it did not do so in the end. Although it offers a point of differentiation for JetBlue, the provision of DIRECTV is associated with extra maintenance costs, problems related to system reliability, and aircraft performance issues as the small screens per seat add to the total weight of the aircraft.

How might a customer equate value to a product or service? Buyers often equate value based on their own perceptions. If the seller of the product has some prestigious or well-known customers, the general public might think the product is better or worth more. If the product is a well-known brand, customers may perceive it to be better than an equally good product made by a lesser known entity.

It is important to note that differentiation strategies are most attractive when the needs and the preferences of the buyers are so diverse that standardization of products is out of the question. There are many different avenues a company can follow in differentiating a product, which has been demonstrated by examining the example of

---

JetBlue Airways' Differentiation Strategy Approach

- Superior service: DIRECTV.
- More for the customer's money: Leather seating and additional leg room.
- Engineering and design performance quality: Excellent safety record.
- Top-of-the-line image: Brand new fleet.
- Technological leadership: Paperless cockpit.
- On-time guarantees: On-time performance.

What are the Competitive Strengths of a Differentiation Strategy?

- **Rival Competitors:** Buyers develop loyalty to brands they like.

- **Buyers:** Bargaining power of larger buyers is not important since other products are less attractive.

- **Suppliers:** May be in a better position to withstand efforts of suppliers to raise prices because buyer is willing to pay premium.

- **Potential Entrants:** Buyer loyalty is a significant entry barrier.

- **Substitutes:** Better positioned against substitutes because the buyer likes the differentiation of the product.

JetBlue. One might look at marketing, sales, and service activities, where the value or perceived value-added is evident after the sale of the product. Or it may be that one can differentiate by changing the method by which one manufactures the product or the way one distributes the product to the end-user.

Keep in mind that a company does not have to add any tangible value to the product as long as the consumer perceives that the product or service offers more value. Perception of value can be as important, or more important, than actual value. Consumers might be more likely to look for this additional value when the buyer is rather unsophisticated. In other words, if the buyer has not done his or her homework and is not well informed of product features and their importance, he or she might tend toward the more well-known product. Likewise, if this is a first-time purchase and it is unlikely that a repurchase will occur in the near future, the buyer may tend to look for the brand name in an attempt to make the best one-time decision possible.

> **When a Differentiation Strategy Works Best**
>
> - If there are many ways to differentiate a product and the buyer perceives these differences to have value.
> - Buyer's needs are diverse.
> - Use of product is diverse.
> - Few competitors are following a differentiation strategy.

Unlike the low-cost strategy, the differentiator is competing on the basis of uniqueness, not cost. Consequently, the differentiation strategy can be expensive. Therefore, it is imperative that the company engaging in a differentiation strategy control costs in order to protect profit margins. But there is a fine line to walk. Although the company must control costs, it has to be sure that the source of differentiation is not lost by trimming too closely.

A good differentiator will distinguish the company on as many dimensions as possible and try to segment the market into many different niches. In addition, it is important to concentrate on establishing brand loyalty. Brand loyalty can be one of the most powerful competitive weapons a differentiator can use. As one would expect, while there are many ways to differentiate a product, some of the most successful sources of differentiation are found through research and development.

Like the low-cost producer strategy, a differentiated strategy can have some competitive shortcomings. Probably the most obvious is that it is very difficult to remain continually unique in the minds of the customers. What also makes this easy or hard is the extent to which competitors are able to imitate the competitive advantage. If the consumer no longer sees the product as being particularly unique, the company has literally lost the competitive advantage because consumers are no longer willing to pay the differential in cost. Also, if price sensitivity ever enters into the picture, differentiation may no longer be an option.

> **What are the Problems with a Differentiation Position?**
>
> - Trying to differentiate on features that either do not lower the operating costs for the buyer or enhance their well-being.
> - Over-differentiating to the point that it exceeds a buyer's needs.
> - Charging a price premium that buyers perceive is too high.
> - Failing to hit the target on what buyers actually consider to be of value.

Niche Strategy

The third Generic Strategy is the niche strategy, where the company pursues either a low-cost strategy or a differentiation strategy but in a very limited segment of the market or to a very limited customer group. Generally, there are three primary ways to segment a market: geographically, type of customer, or product line.

Geographical Niche

With a geographical niche, a company is concentrating on a very defined region or locality. For example, one market with a definite geographical niche is intra-island operations in the state of Hawaii. Two main airlines, Aloha and Hawaiian, concentrate an overwhelming majority of operations on four main islands: Kaui, Oahu, Maui and the island of Hawaii.[5] By focusing on a very specialized geographical area, the companies have been able to build a significant customer base.

Customer-Type Niche

With a customer-type niche market, the company focuses only on a limited or specified customer segment. For example, a company might limit its focus to business travelers exclusively or to wealthy travelers who are more interested in being pampered rather than just getting to another destination. Consider the Concorde service previously offered by British Airways and Air France. There is an enormous volume of daily trans-Atlantic crossings, but in this instance British Airways and Air France (AF) focused on one particularly small segment, the mega-wealthy or extremely urgent traveler who wants personalized ultra-luxury service.

British Airways Concorde Service
Caviar and Champagne in the Stratosphere

- Private Departure Lounge.
- Hand-stitched Connolly leather sculpted seats.
- Commemorative Certificate.
- Meals served on Royal Doulton bone china with rose.
- Freshly laundered linen pillow and blankets.
- Limousine ground transportation arrangements.

---

[5] See <http://www.alohair.com>.

For $12,754 USD round trip, one could travel at twice the speed of sound and travel from New York to London in just under three hours. This was such an exclusive niche market that the Concorde only seated 100 passengers. Shared between Air France and British Airways, there were only a few airplanes, and customers found that the products and/or services were tailored specifically to their own unique needs and preferences.

Did this strategy work for British Airways and Air France? No, but for reasons unrelated to the fact that they had chosen a niche strategy. One of the problems they faced was the very high operational cost of the Concorde. The Concorde was an old aircraft that faced significant operational issues revolving around fuel cost, availability and cost of spare parts, and, in general, the high cost of maintenance. If we compare its financial viability to the financial viability of a brand new aircraft, for example, it is apparent that the Concorde did not perform well at all. Yet a comparison as such would not be an equitable one given the generational difference of the Concorde to an aircraft developed in the 1990s in terms of material used for its construction, engines, avionics, systems, etc. As of 2003, Concordes are museum pieces. A newer niche strategy that some airlines offer focuses on heavily subscribed city-to-city business service, even in long-haul markets, offered by smaller aircraft (B737 or A320 series) that only have business class seats. Speed is not an issue here; instead, convenience and service are the focus.

Product-line Niche

This type of niche strategy focuses on a specific and unique product line. In air service, a very unique niche product-line strategy can be found in Chalk's Ocean Airways. Chalk's is a small firm that operates Grumman Mallard amphibious aircraft in the trans-Gulf Stream market. A typical amphibious operation involves the aircraft taking off from a hard-surfaced runway in Florida in a conventional sense, landing in the destination island's ship channel adjacent to the resort, maneuvering to shore like a seaplane and exiting the water on a ramp by wheels to arrive at a paved area for passenger debarkation. Unconventional barely describes the experience!

As seen in Table 2.2, with its bases in Ft. Lauderdale and Miami, Chalk's flies to resort destinations like Bimini, Walker's Cay and the Atlantis Resort at Paradise Island near Nassau, Bahamas. It is a small tourist market with relatively low frequencies and almost no competition. In fact, it remains the only air service at some of its destinations. It is very adaptable to customer needs and is the expert on the market, having served the region since 1919. The Chalk's example demonstrates a situation where the company elected to compete only on the basis of a narrow or specific niche.

Why would a company want to use a niche strategy? With a niche strategy the company becomes the expert in a particular market. They know everything there is to know about the market and, therefore, can respond quickly to the needs and desires of

Table 2.2 Chalk's Timetable[6]

Paradise Island to Ft. Lauderdale							
Flt. #	Depart	Arrive	Scheduled	Flt. #	Depart	Arrive	Scheduled
OP 501	8:00am	9:05am	Daily	OP 502	9:35am	10:40am	Daily
OP 507	10:10am	11:15am	Monday, Thursday	OP 508	11:40am	12:45pm	Monday, Thursday
OP 503	11:15am	12:20pm	Daily	OP 504	12:50pm	1:55pm	Daily
OP 515	12:35pm	1:40pm	Friday, Sunday	OP 516	2:10pm	3:15pm	Friday, Sunday
OP 505	2:25pm	3:30pm	Daily	OP 506	4:00pm	5:05pm	Daily

that market segment. Because of this flexibility, and the fact that the company is not trying to be all things to all people, they can respond and adapt much quicker than a multi-segmented firm. If a firm is successful in quickly and accurately anticipating the needs of the market, they are more likely to develop brand loyalty, which then acts as a buffer to other entrants into the market. Sometimes a niche strategy is referred to as a boutique strategy because the company often gains the total loyalty of a small following and generally serves as the only supplier for that market.

Cost or Differentiated Niche Strategy

A company can pursue a low-cost or a differentiated niche strategy, a combination of the Generic Strategies. When the company chooses to compete on cost, it must make sure that the cost of producing the product or providing the service is significantly lower than the industry cost leader in the specific or narrow segments of the market. For example, it is much more costly for a multi-segmented airline to make short-hauls than it is for a boutique airline that concentrates on point-to-point short trips. This is exactly what Continental Airlines found when trying to compete with Southwest Airlines. In 1993, Continental started Continental Lite, a smaller no-frills regional airline operating under the umbrella of Continental. But Continental soon found that it could not be all things to all people and keep the costs low enough to be profitable. The niche strategy of Southwest was too much to compete with for the full-service airline because Continental could not sustain a low-cost advantage with short-haul flights. A cost-niche advantage can also be successful if the niche company can produce custom products or services that are difficult for a multi-segmented company to copy given a concern for cost efficiencies.

On the other hand, the differentiated niche strategy allows the company to employ the exact same type of differentiated strategies that a multi-segmented firm would use,

---

[6] See <http://www.chalksoceanairways.com/sub1k.htm>.

but it does so by serving only a small segment of the market. Again, the company competes in the marketplace by providing a product or service that is unique; however, this time the uniqueness is directed at a much smaller segment. It is expected that this segment will be willing to pay extra for the unique qualities. When the differentiated niche strategy is employed the company can target even more precisely the needs, desires and perceived expectations of the market it is serving.

What are the Competitive Strengths of a Niche Strategy?

- **Rival Competitors:** Rivals do not have the ability to specialize enough to meet the needs of the target market.

- **Buyers:** Because the company focuses so closely on the unique needs of the market, the bargaining power of the larger rival firms is diminished.

- **Potential Entrants:** Because the niche competitor generates a core competency for that market, it acts as a barrier.

- **Substitutes:** Core competencies serves as a barrier to substitutes.

With a successful niche strategy, the company does a better job of serving the buyers in the target market than its larger rival firms. For this to work, the company must choose a market niche where buyers have distinctive preferences, special requirements, or unique needs. If this market segment is an important segment to the welfare of a larger, more financially viable company, the niche firm might not be able to compete.

When a Focus Strategy Works Best

- When it is cost prohibitive for a larger, more diverse rival to serve the specialized needs of a smaller market.
- No other firms are concentrating on this market or include this market as a major market segment.
- The firm does not really have the resources to be multi-segmented.
- There are many segments in the industry.

What makes a particular segment of the market attractive to the niche player? First, the segment has to be large enough that there is a good potential for profitability.

Likewise, the segment cannot be in a stagnated industry. Specifically, there has to be reasonable potential for growth in the segment. Given this, the niche firm has to have sufficient resources already in place to serve this market segment fully. This is an important issue. The niche firm must come into the market segment in full force and immediately try to gain customer loyalty. A slow start might give way to competitors entering in much the same way, which might then begin to erode the niche firm's position of power.

What are the Problems with a Niche Position?

- The niche firm is operating on a small scale, which often makes it difficult to keep costs low.
- Technological changes can affect the niche market because smaller firms might find it difficult and even cost prohibitive to implement.
- There is always the chance of a larger competitor, and more lucrative, infringing on the niche market by offering similar products and service.

A niche strategy can be a very powerful strategy in a fast-growing industry with fast growing segments. However, is it important that the market be big enough to be profitable yet small enough that it is only of secondary interest to any large competitors. In addition, this strategy will not work with commodity products because for this strategy to be successful, the buyers must require specialized expertise or require some kind of customized product or service attributes in the segment on which a niche strategy is focused.

A very interesting niche strategy is one offered by providers of what is called "fractional ownership." It is similar to buying a "timeshare," in the sky. In the current economic environment where margin, efficiency, utilization, tax implications, maintenance and economies of scale rule conventional wisdom, some companies are moving away from either using the services of airlines or owning and operating their own corporate fleets and, instead, are moving toward sharing business jets. An example of a company active in this business segment is NetJets, a Berkshire Hathaway company, who pioneered the fractional ownership model in 1986. Several other firms such as Flight Options, Raytheon Travel Air, Indigo, Avolar and Delta Business Jets have also entered into this marketplace. Fractional ownership companies provide for thousands of passengers to fly to destinations around the world each day.

Fractional ownership programs include several different aircraft types, ranging from small cabins (a few seats) to large cabins (18+ seats), with flight ranges up to 7,500 miles. Some of the fractional ownership companies provide their customers with the ability to change plane size when needed. Fractional ownership providers establish maintenance procedures with each manufacturer to assure that their aircraft are maintained according to regulations. In terms of aircraft guarantee, NetJets, for example, guarantees its customers to have an aircraft at their requested location within four to 12 hours of their call depending on interest size and aircraft type. Generally 48-hours notice is required for "peak period days." NetJets is also an international operator of fractional ownership, providing interchange between their U.S., Europe and Middle East programs.[7]

Best-Cost Producer Strategy

While not one of Porter's initial Generic Strategies, the best-cost producer strategy is a combination of Porter's cost and differentiated strategies. In other words, many companies, particularly companies in very competitive industries, emphasize both low-cost and differentiation. How do they do this? They might make an upscale product or provide a superior service at costs lower than the competition. By doing this, they can sell the product or service at a lower price than their rival and, thus, give the buyer a better product with more value for the money they are spending. In this situation, the company tries to meet or exceed the buyer's expectations on both the product attributes as well as the price.

With a best-cost strategy it is obvious that the competitive advantage comes from being able to match or exceed the key attributes offered by rival firms but, at the same time, providing the product or service at a price they cannot match. A best-cost producer can out-compete both a low-cost provider and a differentiated provider in situations where buyer diversity makes product differentiation a necessity and the buyers in the market are sensitive to price.

The Miles and Snow Typology

Strategists Raymond Miles and Charles Snow chose a different route in their attempt to categorize competitive strategies.[8] The basic principle of their strategic typology proposed that competing firms within an industry exhibit four basic patterns: defender, prospector, analyzer, and reactor. The assumption is that the firms compete differently in an industry because they view their environments on an idiosyncratic basis and make resource allocation decisions based on these views.

---

[7] See <http: //www.netjets.com/aboutnj/>.
[8] Raymond Miles et al., "Organizational Strategy, Structure, and Process," *Academy of Management Review*, 3, no. 3, 1978: 546–563.

Defender

An organization that follows the defender orientation tends to have a narrow product market. This type of firm will try to create and maintain a niche in the marketplace with a limited range of products and services and a narrow technological base. A defender firm does not try to search outside of its "comfort zone," and, therefore, becomes highly dependent on its narrow product/market area.
How does a defender firm protect its domain? Usually a defender will use lower prices, higher quality and superior delivery. The organizational structure of a defender firm is generally very rigid, centralized and formal. There is little flexibility in the structure of a defender firm.

What are the problems with being a defender? It is difficult to sustain a competitive advantage when a company relies on price, quality, and delivery to protect its domain all of these factors are easily copied and hard to sustain for a long period of time.

Prospector

A firm that follows the prospector typology is continually in search of new markets. Unlike the defender firm, a prospector has a broad and flexible product/market domain and a broad technical base. A prospector is the creator of change in the marketplace. They respond quickly to early opportunities and are usually the first to enter into a new product/market arena. The primary concern for a prospector company is product/market innovation.

The organization structure of a prospector firm is flexible, with a low degree of centralization and routinization as well as a low degree of formalization. Prospector firms have open communications, both vertically and horizontally.

Analyzer

An analyzer firm is somewhat of a cross between a defender and prospector. Firms characterized as analyzers tend to maintain a stable and limited domain. They move into new markets and products cautiously and only after these markets and products have been proven viable by the prospector in the marketplace. Analyzers take the promising ideas of prospectors and successfully market them. They seek flexibility as well as stability. Their organizational structures are stable yet flexible enough to respond to changing domains.

Reactor

A reactor firm does just what it says – it reacts. It has no long term objectives or strategies. A reactor firm has no consistent pattern of behavior. It is passive in dealing with most issues. Reactors do not attempt to maintain a defined product/market domain nor are they alert enough to respond to environmental opportunities.

A Fresh Perspective on Competitive Strategies

Porter's Generic Strategies and the Miles and Snow typology have laid the groundwork for studying competitive positioning for the past three decades. But will these rather simple strategic positions work in today's highly competitive, global workplace? One of the problems is sustainability. It is not enough for a company to operate within a low-cost structure; in a highly competitive environment cost is the norm. One cannot compete effectively in the marketplace if cost is not a primary consideration. Likewise, it is not enough to offer high quality, because quality is not even a bargaining position today. The quality requirements many companies place on their suppliers today makes quality a non-entity. If the quality is not there, the company will not buy. It is as simple as that.

Porter argued that positioning, which is the heart of strategy, is too static for today's dynamic, technologically advanced marketplace.[9] In other words, rivals can quickly copy any market position and, therefore, any competitive advantage is only temporary. How does a company compete in today's environment then? A company must have core competencies and must understand the concept of trade-offs in order to build long lasting strategies that will take them into future generations. Core competencies, a concept developed by Hamel and Prahalad, are the organization's major value-creating skills and capabilities.[10] These skills and capabilities are shared across multiple product lines and multiple businesses. This is precisely what makes it a core competency – the sharing. For instance, consider the skills and talents of an individual person. What is it that he or she does extremely well, something better than anyone else? This talent could be called their core competency. In Chapter 3 we will study the concept of core competencies and how they can be used to gain a sustainable competitive advantage.

What are trade-offs? Choosing a unique position is not enough to guarantee a sustainable advantage. A sustainable strategic position is not possible unless there are trade-offs with other positions and they occur when activities are incompatible. In other words, more of one thing means less of another. For instance, an airline can choose to serve meals or it can choose not to, but it cannot do both without bearing major inefficiencies.[12] Serving meals slows down turnaround time and adds expense. Trade-offs force a company to choose between one activity over another. In Chapter 3 we will talk in more depth about trade-off and their relationship to competitive sustainability.

---

[9] Porter, 61–78.
[10] Gary Hamel and C.K. Prahalad, "The Core Competence of the Corporation," *Harvard Business Review*, 68, no. 2, 1990: 79–93.
[11] Mahmood S. Bahaee, "Strategy-Comprehensive Fit and Performance," *Australian Journal of Management*, 17, 1992: 195–215.
[12] Porter, 61–78.

Adapting the Miles and Snow Typology to the Regional Airlines[11]

The 1978 deregulation of the U.S. airline industry created an almost immediate opportunity for short-haul carriers to enter the market – a market that was previously thought to be saturated. Deregulation suddenly opened the flood gates to a dynamic growth market. Prior to deregulation, the competition among carriers was minimal due to the high degree of regulations and the competition between regional and national carriers was completely non-existent. Deregulation eliminated fare controls and just about abolished any barriers to entry. Carriers were now permitted to select their own routes. Many new, entrepreneurial airlines entered the market as a result of deregulation. But these new opportunities were in a sense thrust on the industry – companies did not go out and proactively seek these changes.

Prospector firms were in a position to respond to the dramatically altered environment while the other firms had to adapt a more flexible, entrepreneurial behavior, or they would risk acquisition or failure. But there were few prospectors in the regional airlines. Many blamed this on the highly capital intensive nature of the industry, which limited flexibility.

What became obvious was that there were a high number of reactor firms that emerged after deregulation. Most of the regional airlines were caught off guard. They had never had to articulate a strategic plan or set long-term objectives. They were satisfied with short-term profitability because they saw the new industry as affording many new opportunities. What they were not prepared for was how competition would affect their market position.

A Summary of Competitive Strategy

The framework established by Porter with his Generic Strategies is by far the most popular approach to assessing how organizations compete in the marketplace. While these Generic Strategies are important to understand, operating only in a low-cost leadership position or only in a differentiated position is probably not entirely feasible today. As we noted, keeping costs low is a necessity to succeed in any business at the present moment. Likewise, if a company is differentiated and unable to keep costs low, the likelihood for long-term success will be gravely diminished.

Companies in the twenty-first century must rely on their core competencies to take them into the future. If this success in the present cannot move companies into the future and provide a platform on which to build next generation products and services, then any advantage companies have at the present moment will be quickly eroded. Companies who have learned to provide quick, efficient, on-time service while still keeping their costs low in a manner that is difficult to imitate, like Southwest Airlines, will find future sustainability against a counterpart that is still trying to provide full service on a larger scale. Part of what it means to be competitive is to be flexible enough to respond quickly and proactively to the changing needs of the environment.

No matter how competitive strategies are categorized, it is important to realize that the company's competitive strategy must exploit core competencies in order to obtain and sustain a competitive advantage. Without a competitive advantage, it will be extremely difficult to compete successfully in today's global marketplace.

Chapter 3

The Essence of Competitive Strategies

> The development of theories of bounded rationality needs an empirical basis.
>
> Reinhard Selton[1]

> Radical change raises the level of complexity.
>
> Robert Saloman[2]

It might be surprising to know that most companies do not have a competitive advantage. Does this mean that they are not doing things well? Not necessarily. They may have good products and good relations with their customers. In fact, they may have some excellent capabilities, or they may have features on their products that no one else has, but these alone do not add up to a competitive advantage. Realistically, how many companies can truly dominate a market or a marketplace? Not many.

However, maybe a company does not have to dominate the marketplace to have a competitive advantage. This may seem like a contradiction, but consider the situation this way: maybe a competitive advantage does not have to be as black and white as previously believed. Traditionally, the focus has only been on the outcome, but a competitive advantage can also be found within the process. Maybe it is not necessary to "own" the marketplace to be the most competitively savvy company in the arena.[3]

In the past, people believed that bigger was better, but how easily can one respond to changes in the marketplace if the company is bogged down in the bureaucracy that generally accompanies big business? Today, we know that it is hard to maintain a competitive advantage unless a company has flexibility. The less investment in brick and mortar the easier it is to move or to redirect. This is not to say that maintaining a competitive advantage means the company must curtail any investment in brick and mortar, nor does it mean that a large company cannot be competitively successful, because they can. What has been made perfectly clear – particularly in the airline industry – is that small resource-limited competitors can successfully challenge the market.

---

[1] Rheinhard Selton, "Factors of Experimentally Observed Bounded Rationality," *European Economics Review*, 42, 1998: 414.

[2] Robert Soloman, *Competitive Intelligence: Scanning the Global Environment*, London: Economica LTD, 1999, 1.

[3] Bruce Chew, "The Geometry of Competition," Cambridge, MA, 2000, see <http://www.monitor.com/binary-data/MONITOR_ARTICLES/object/101.pdf>.

Flexibility and Competitive Advantage

At present, airlines need to formulate a new attitude. They cannot simply exist within the industry; instead they need to shape the industry in a manner that suits them best. With all the turbulence that has occurred in the industry over the past few years, even before September 11, 2001, it appears that the old proven "attitude" is being challenged. Tables 3.1 and 3.2 illustrate the differences between historical operational and service offerings in the airline industry and new operational and service offerings.

Table 3.1 Historical Operational and Service Offerings

- **Fleet**
  Wide breadth of fleet to fill all niches. For example, TWA had a fleet of B747, B767, L1011, B757, B737, MD80, DC9 and B727 aircraft. TW Express had J41, J31, and ATR72.
- **Seating Configuration**
  Three levels of seating: first class, business class and economy class.
- **Revenue Management Systems**
  Complicated revenue management schemes in an effort to capture higher profits.
- **Airports**
  Customers flew to the airport closest to the city center.
- **Meals**
  Meal service in first class and economy.
- **First Class Service**
  Bigger, more comfortable seats.
- Ticketing and Interlining
  Paper ticketing and wide use of IATA services for interline tickets.

In "Rebuilding the Corporate Genome: Unlocking the Real Value of Your Business," the authors suggested that the day of the integrated company, organized around strategic business units, is over. Instead, they say that a sustainable competitive advantage can be obtained from a different, more flexible model. The new model they propose will be a collection of separate businesses, organized around the different business capabilities, of which business: manufacturing, assembly and distribution, development, design and even branding are all potential examples.[4] Hamel and Prahalad said that if a company excels in one area or one competency, they should organize the entire business around

---

[4] Johan C. Aurik, Gillis J. Jonk and Robert E. Willen, *Rebuilding the Corporate Genome: Unlocking the Real Value of Your Business*, Hoboken, NJ: John Wiley, 2003, 3–19.

this competency.[5] What does this mean? Some companies, such as Ericsson and Motorola, have decided to outsource the manufacturing of goods and concentrate on their competencies of design and marketing.[6]

Table 3.2 New Operational and Service Offerings

- **Fleet**
  Consolidated Fleet: most of the low-cost carriers have common type fleets that help lower operational costs by commonality of training, parts, and service. Delta, Continental, and American Airlines are examples of airlines looking toward cost savings in uniformity. CO flies 737, 737NG (New Generation), 757/767, 777, and MD80 series. Delta is transitioning to 737, 737NG, 757/767, MD80 series and 777. American Airlines is changing to 737NG, MD80 series, 757/767 and 777.
- **Changes in Seating Configurations**
  Addressing what customers are willing to pay for and accept: "cattle-car" all economy (e.g. Southwest Airlines); full service (e.g. Singapore Airlines on long-haul flights); Business First mix (e.g. Delta long-haul flights); First, Business, Economy Plus and Economy (e.g. British Airways on long-haul flights).
- **Revenue Management Systems**
  Some airlines are focusing on less complicated, more manageable fares (e.g. Song, Southwest, JetBlue, Ryanair and EasyJet). Some legacy carriers are doing this as well (e.g. Air Canada).
- **Airports**
  Travelers now fly to outlier cities for lower fares (e.g. Providence, RI; Manchester, NH; JohnWayne/Burbank/Ontario/LongBeach, CA; Islip, NY; and Birmingham, AL). Airlines can get better turnaround times, lower landing tariffs, less congestion, and more city/airport development perks at these airports.
- **Meals**
  In Economy, no meal service or customers have the option to buy name-brand food.
- **First-Class Service**
  Sky Suites with fully reclining chair beds and on-demand video/movies.
- **Ticketing and Interlining**
  E-tickets and less reliance on interlining to cut costs and maintain better control of inventories.

---

[5] Gary Hamel and C.K. Prahalad, "The Core Competence of the Corporation," *Harvard Business Review*, 68, no. 2, 1990: 79–93.
[6] Aurik, Jonk and Willen, 3–19.

Core Competency

In Chapter 2, core competencies were referred to as the major value creating activities of the company. Hamel and Prahalad called them the collective learning of the organization. Other experts view core competencies as that part of the value chain which make a unique and value-added contribution to the business. Although there is no one correct definition, they all connote the essence of what is considered a core competency. In general, if a company does something well it should exploit it and build the business in a manner that will highlight this competency. Therefore, when one talks about core competency and competitive advantage, it is important that the competency comes from within the organization and possibly within the value chain.

How Does One Recognize a Core Competency?[7]

How does a company identify its core competency? One company set out to try to find its competitive advantage. They called in some competitive advantage experts to try to help them in their quest. The experts led top management through a series of group exercises and the result was that they had a bigger mess than when they started, because they could not distinguish what was core as opposed to non-core, or what was a competency as opposed to something that they just did relatively well.

This example leads to an interesting observation: core competencies are essential to competitive advantage, but people seem to have a difficult time figuring out what is a core competency. Maybe it is because many people are looking for a tangible thing to be the core competency. Core competencies can cover skills and processes, like engineering, production and even corporate identity. A core competency can be a combination of skills and knowledge, but it is something that is embedded in the organization, and because it exists the company is able to reach to new heights, even world class operations.

How can an executive determine if his or her company has a core competence? Four key questions should be asked (see box).

Let us evaluate each of these questions.

Are Our Skills Truly Superior?

A core competency requires the company to be the best, or nearly the best, at the chosen competence. A competence-based strategy cannot merely be grounded on a skill that is important to a business or something that customers like. The company

---

[7] Kevin P. Coyne, Steven J.D. Hall, and Patricia G. Clifford, "Is Your Core Competency a Mirage?," *McKinsey Quarterly*, 1, 1997, see <http://mckinseyquarterly.com/article_abstract.aspx?ar=186&L2=21&L3=37>.

> **Questions to Ask to Determine Core Competencies:**
>
> 1. Are our skills truly superior?
> 2. How sustainable is the superiority?
> 3. How much value can the competency generate in comparison to other economic levers?
> 4. Is the competency integral to our value proposition?

has to be better at this competence than all others or at least the vast majority of its competitors.

How Sustainable is the Superiority?

The first question a manager wants to ask is, "How long will it take the best positioned competitor to copy the company's competency?." Ease of imitation is a good indicator of just how unique or rare the competence really is. The speed in which it takes a competitor to develop the same competency coupled with how difficult it is for a competitor to understand the source of the competency are important in determining ease of imitation. How does a manager determine if the competence is sustainable? He or she has to look at the rareness of the competence as well as the source of the competition.

Rareness A manager can determine rareness by comparing the company's competence to those in other firms in a variety of industries. Keep in mind that rareness does not come from examining only one industry.[8,9] The way in which a world-class hotel treats its customers may be quite applicable to the airline industry, yet it may be difficult to reproduce. The time it takes to copy a competency is also important to rareness. Even if the competency can be reproduced, if properly implemented, it may take months or years for the competitor to develop the necessary support systems, train the appropriate individuals and establish the necessary policies to implement the competence.

Source Whether or not the source of the competition can be understood by competitors depends upon the source of the competition. Consider the hotel example, again. If the hotel's competitive advantage in dealing with customers is an essential part of

---

[8] Michael E. Porter, "What is Strategy?," *Harvard Business Review*, 74, no. 6, 1996: 61–78.
[9] Jay Barney, "Firm Resources and Sustained Competitive Advantage," *Journal of Management*, 17, no. 1, 1991: 99–120.

the corporate culture, sometimes the culture of the organization cannot simply be reproduced. Thus, the strategy may be sustainable. Also, the number of aspects involved in the competency also has an effect on whether it can be sustained. If there are a lot of intermingling components to the strategy, it might be extremely difficult for anyone to reproduce.

How Much Value Can the Competency Generate in Comparison to Other Economic Levers?

Being the best at something does not offset other disadvantages. For example, if the industry is on the decline, no matter how good the strategic competence, it is unlikely the company can overcome the pressures of a declining industry. The competency must be stronger than any other external and internal forces.

Is the Competence Integral to Our Value Proposition?

Investment in the core competence must be aligned with what will be rewarded in the marketplace. A company does not want to invest in a competency of customer service if it does not first position itself in the marketplace as the best service provider.

Turning Core Competencies into Sustainable Competitive Advantages[10]

The last question, "Is the competence integral to our value proposition?" is the key to turning the competency into a sustainable competitive advantage. If it is not looked upon favorably in the market or if the value-added is not of interest to the customer, the competency will not lead to sustainability. What is really meant by a sustainable advantage?

Sustainable Competitive Strategies

The goal of strategy is to have a sustainable competitive advantage, but what is an advantage and when does a manager know when there is an advantage? These are the pressing questions in business today.

Strategy is about being different, but not in the same way as the Generic Strategy of differentiation. For a competitive strategy to be sustainable, company managers have to choose a different set of activities from the competitors and deliver them in a way that creates a unique value. It requires positioning and fit. Does this sound

---

[10] Bruce Chew, "The Geometry of Competition," Cambridge, MA, 2000, see <http://www.monitor.com/binary-data/MONITOR_ARTICLES/object/101.pdf>.

remotely like core competencies? It should! But even this is not enough. It also requires fit and focus.

Fit and Focus[11,12]

It almost seems unfair that being better than the competition is not enough to have a sustainable competitive advantage, but the reality is that fit and focus are necessary to a sustainable competitive strategy.

- *Fit*: It sounds really easy, but it is not. A round peg will not fit into a square hole and, therefore, it is immediately evident there is no fit. But fit goes way beyond this. Some companies feel that if they have a competitive advantage in one market then it will work in another. This is not necessarily true; the fit was in the first market, not the second. Sometimes it is exactly what made that company successful in one market (e.g. the full-service airline market) that makes them unsuccessful in another. We will examine a good example of this problem with Continental Lite.
- *Focus*: It seems like everyone wants to be full service, to be everything to everyone. However, it is difficult to be outstanding when the company is focusing its attention in many different directions. To be competitive, a company needs to focus on what they offer to the market.

Can any Strategic Position be Copied?[13]

It might seem that any competitor could copy any market position; however, this is not always the case. Consider what happened in the airline industry. It seemed to Continental that Southwest could be copied. Continental saw how successful Southwest was operating solely short-haul runs. In response, Continental took a straddling position. A straddling position means that a company tries to match the benefits of a competitor's position while still maintaining its existing market position. In Continental's case, the airline wanted to remain a full-service airline while at the same time trying to mimic the services and success of Southwest. It did this under the name of Continental Lite. Within the parameters of Continental Lite, meals were eliminated as was first-class service. Compared to Continental's full-service component, Continental Lite had increased frequency of departures, lower fares, and shorter turnaround at the gate, just like Southwest. However, the full-service component of Continental Airlines operations remained the same. They still used

---

[11] Ibid.
[12] Porter, 61–78.
[13] Ibid.

travel agents, baggage checking and seat assignments as well as a mixed fleet of airplanes. Were they successful?

One of the important things to remember about sustaining strategic positions is that a position is not sustainable unless there are tradeoffs with other positions. In other words, the activities of the full-service Continental operations were incompatible with the Continental Lite operations. In the long run, Continental found it could not be everything to everyone in their original market and still compete against Southwest in the no-frills market.

When are Tradeoffs Important?[14]

As previously discussed, it is not enough to have a unique position or a unique way of doing business because this still leaves the company susceptible to imitation by competitors. A company can choose to reposition itself and go head-to-head with the competition by offering a level of service that is just a notch above what they are doing. However, most companies choose the straddling strategy, they add new features or services to what they are already doing. Why did this fail for Continental? A company cannot sustain a competiive position without tradeoffs. Continental Lite was not compatible with Continental's other operations. An airline can be no-frills like Southwest or offer full-service activities, but it is difficult to do both.

Tradeoffs are Essential to Sustainability[15]

Essentially, tradeoffs force a company to make choices between services. They require companies to limit, purposely, their offerings. Tradeoffs arise for three reasons. The first reason is that the new image or brand that the company is trying to project is inconsistent with its former image. If a company is known for one type of service or value, it can become confusing to customers if another type of service or value is offered; the new image might even undermine the reputation of the company. Continental Airlines was a full-service airline and Continental Lite was a different kind of "value" for the company.

The second reason that tradeoffs arise is from the activity itself. A new activity often requires different skills, equipment, management systems, or employee attitudes.

---

[14] Ibid.
[15] Ibid.

If a company must adjust to meet the needs of the new market segment, it is likely that the original customer will suffer. Interestingly, it is often because of the company's inability to adjust to the new activity that a tradeoff becomes necessary. For instance, if originally the company offers full service and then begins to offer no-frills service, it is possible that the employees involved are not trained for the new service. In other words, it might be difficult for these employees to offer no-frills service because their initial training and company mindset were different.

Finally, tradeoffs come about as a result of limits on internal coordination and control. When a company operates in one market and mode of operation, such as Southwest Airlines, the organizational priorities all revolve around this means of operation. When a company tries to be everything to everyone, such as Continental Airlines attempted to do, the organizational priorities become clouded. Continental Airlines could not successfully implement a new activity without tradeoffs, failing to straddle full and no-frills service.

Case Illustration: Institutionalizing Competitive Advantage: Southwest Airlines' Unique Advantage[16]

Southwest, headquartered in Dallas with its base at Love Field, has a history of expanding into new markets where it is welcomed by consumers for causing airfares to drop. Southwest's core business strategy has been to be a low-cost, no-frills alternative to the legacy airlines. The airline was founded by Rollin King and Herb Kelleher, who decided to start a "... different kind of airline, with one simple notion: if you get your passengers to their destinations when they want to get there, on time, at the lowest possible fares, and make darn sure they have a good time doing it, people will fly your airline."[17]

To save on costs, Southwest uses a standardized fleet of the Boeing 737 family of aircraft. In addition, it flies into lower traffic (secondary) airports that charge lower fees and provide them with chances for faster aircraft turnaround times (because of lower traffic density) but which are near major urban centers. In several cases, Southwest has also been welcomed by airport authorities who offered incentives to start operations from their airports in an effort to stimulate economic growth. They have specialized largely on short-haul routes, particularly during their early history, marketing themselves as an alternative to other modes of transportation, for example cars. It has a cooperative relationship with the employee unions, a fact that boosts productivity through better morale but also lower operating costs through friendlier pilot contracts. Southwest was also a pioneer of ticketless travel. They also use direct channels of ticket distribution to lower costs and avoid travel agent commissions. All

---

[16] See <http://www.southwest.com/about_swa/airborne.html>.
[17] Ibid.

of the above factors combined have allowed Southwest to be consistently profitable since its inception in 1971.

In 2003, Southwest was the sixth largest U.S. airline and ninth largest airline globally as measured in Revenue Passenger Kilometers. As of the beginning of 2004, Southwest provided high-frequency, "point-to-point" air transportation between nearly 60 locations in the United States (58 cities and 59 airports in 30 states). Its expansion strategy complements its business model. Southwest tends to concentrate on secondary airports, located near major urban centers, and tries to stimulate traffic with its inauguration of new service rather than diverting traffic from established routes. For example, its move into the New England area demonstrates that Southwest targets secondary airports in major urban centers. In 2003, it began service to Providence, RI, and more recently to New Hampshire, allowing Southwest to bracket Boston's Logan Airport from both the North and South, therefore providing access to a major urban market.

The Southwest brand provides value to consumers, investors and the industry. Southwest has achieved a reputation as the most efficient operator of the B737 family of aircraft through its specialization in the utilization of the aircraft. Since the accident of Valujet Flight 592 (DC9) in the Florida Everglades on May 11, 1996, several questions have been raised regarding the safety performance of low-cost airlines. Southwest has been able to quell these concerns with its excellent record of safety in addition to its aforementioned record of operational efficiency.

It appears that Southwest is an extremely well-managed company that has focused on strong human resource management and the creation of a unique company organizational structure as a key means of achieving its sustainable comparative advantage. In 1997 and 1998, Southwest was named "best company to work for in America" by *Fortune Magazine*.[18] According to its annual 10K filing with the Securities and Exchange Commission (SEC), Southwest Airlines earned profits of $241 million USD in 2002, while the eight largest carriers combined lost $11 billion USD. Southwest has recorded 30 straight profitable years in a volatile industry, which is notorious for "razor thin" profit margins. Its good financial performance can be attributed to its efficiency and the conservative internal financing of its expansion. The team spirit and productivity of its employees and the loyalty of its customers are also worth mentioning. Salary expenses remain low because the airline supplements salaries with stock options as part of compensation. Further it has more flexible work arrangement than competitors.

Despite these positives, the reality is that Southwest operates in a competitive industry facing unpredictable fluctuations in the economy and fuel prices. Southwest Airlines built its strategy on providing low-cost convenient service on short-haul routes. Can this be imitated? Of course. But it is not just this that has made them successful. Certainly, Southwest concentrated on one segment and is not trying to

---

[18] See <http://www.southwest.com/about_swa/press/factsheet.html#Recognitions>.

be everything to everybody; however, it is also the entire package that Southwest has created that has made it hard to beat.

Southwest made sure that its low-cost strategy was integrated with all aspects of its firm's existence. For example, no-frills means that personal paper tickets are replaced with laminated boarding cards issued at the gate. The passes are numbered and given out on a first-come first-serve basis to encourage passengers to show up early, as those with the first passes are allowed to board first and claim space in the non-reserved cabin. Boarding cards are reused by the gate agents and are easy to process, making it easy to determine which passengers are on board. This allows for the fastest boarding in the industry.

Flight attendants wear polo shirts and khakis or shorts and, therefore, uniform costs are minimized. The casual attire also seems to fit well with Southwest's informal image and focus on the leisure traveler. Promotional efforts match the client. Employees are fun and friendly to their clientele and encourage travelers to take Southwest instead of driving to their destinations. As a marketing ploy in the early days, Southwest provided a complimentary 1/5th liter of whiskey to customers as a means of seducing them away from Braniff and American.

Some aircraft are painted in colorful livery to attract leisure travelers. One plane is painted with the Sea World mascot Shamu, another with the California state flag and another with the Arizona state flag.

Are Southwest's skills superior and its position sustainable? The future will be the ultimate judge of this, but past experience has shown that Southwest has been able to maintain its profitability in difficult financial times, to contain its costs, and to keep its workforce motivated. If it is able to sustain these cornerstone advantages, then it should be able to maintain its strategic positioning.

Chapter 4

The External Environment

> Every other start-up wants to be another United or Delta or American. We just want to get rich.
>
> Lawrence Priddy[1]

> I really don't know one plane from the other. To me they are just marginal costs with wings.
>
> Alfred Kahn[2]

> There are a lot of parallels between what we're doing and an expensive watch. It's very complex, has a lot of parts and it only has value when it's predictable and reliable.
>
> Gordon Bethune[3]

> A recession is when you have to tighten your belt; depression is when you have no belt to tighten. When you've lost your trousers – you're in the airline business.
>
> Sir Adam Thomson[4]

Effective strategies make it possible for a company to compete successfully in the aviation environment, but good strategies do not come without hard work and detailed planning from the CEO. A well-thought out strategy is the output of a thorough analysis of the environments affecting the company. Thus, determining the appropriate strategy should come directly from an assessment of the external environment and the company's internal environment. In this chapter, we will discuss the external environment. The external environment encompasses all of the relevant forces outside of the company's control. While one company cannot have a major effect on these forces, changes in external environmental forces can have a significant effect on the way the company does business. In other words, a company could be forced to change its business model or strategic direction based solely on a change in one of these relevant forces. Furthermore, just as there is a micro and macro analysis in economics, a company is affected by both macro and micro-environmental forces.

---

[1] See <http: //www.skygod.com/quotes/airline.html>.
[2] Ibid.
[3] Ibid.
[4] Ibid.

The Macro External Environment

The macro external environment consists of such things as the economy, unemployment levels, inflation, the demographics of the population, governmental legislation and regulations, political power and stability, technology and even the values and lifestyles of the society at large. The macro environment is shown in Figure 4.1. A manager might think that these environmental forces are well beyond the scope of the company's influence; however, failure to keep a close watch on these forces could cause an airline to be ill-prepared for the future.

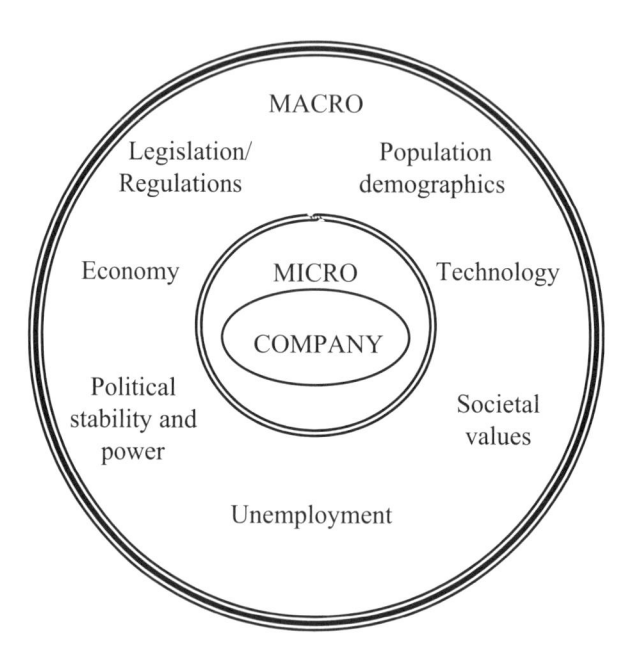

Figure 4.1 The Macro External Environment

Let us begin our examination of the macro external environment with the economy. During times of economic prosperity, as was the case during the late 1990s, people have more disposable income. It is during these times when one might see an increase in leisure time activities, such as family vacations and weekend getaways. Economic prosperity means good times for airline-related companies. Conversely, airlines and related businesses suffer greatly during downturns in the economy because disposable income is less prevalent and the general public tends to stay closer to home. It is at times like these that the business traveler, rather than the leisure traveler, becomes the focus for many airlines.

Unemployment is another macro environmental factor. During times of full employment, the labor market is a seller's market. In order to attract good employees, a business might have to pay top dollar, or they might find themselves having trouble hiring enough qualified employees. In general, full employment goes hand-in-hand with economic prosperity, which can mean possibilities of expansion for the company. However, full employment might also cause an airline to curtail expansion plans because of the lack of qualified available labor. Inflation, like unemployment, can cause problems for the businesses as well. During times of high inflation, the leisure travelers tend to curtail travel and businesses tend to put plans for expansion on hold.

Governmental legislation and regulations can have some of the most significant influences on a company's present and future operations. When the airline industry was deregulated in the United States in 1978, the commercial airlines were thrown into a completely new competitive environment. This new competitive environment was based on price, frequency and networks rather than in-flight service alone, as in the pre-1978 era. The Airline Deregulation Act changed the airline industry forever. Sometimes regulatory changes have merely procedural affects on operations. For example, prohibited Area P-51 is a parcel of restricted airspace over George W. Bush's ranch in Crawford, Texas, created after his election in 2000. All flights were forced to detour around this airspace. In other times, regulations can create costly changes to present operations. For instance, Naples (Florida) Municipal Airport recently banned all Stage 2 Jet operations to the airport. Operators are now required to retrofit older jets with costly hush kits to soften their "noise-print." Another example is the regulated changes in airport security following 9-11, which, while important, were costly and caused many airports to reassess where monies were spent. Legislative changes may open up new markets to some and close markets to others. Any such change would force the businesses involved to reassess and perhaps redirect their strategic focus.

Political power can also be a key element in countries where the airlines remain state-owned. In this context, the strategic direction of the airlines and their strategic viability is often in the hands of the politicians. Olympic Airlines (formerly a part of Olympic Airways) is a prime example of this. The Board of Olympic Airlines is appointed by the Greek government, a practice that many would suggest has caused significant instability for the airline and contributed to its weak competitive position. Olympic Airlines was also used by politicians as a means of returning political favors to their constituencies.[5]

Political instability, on the other hand, can alter general domestic travel throughout the world. For 40 years, air travel was prohibited between the U.S. and Cuba because of the U.S. embargo against Cuba imposed shortly after Fidel Castro came to power. As

---

[5] An extensive study of this, in Greek, has been conducted by Giannis Lainos, *Fakelos: Olympiaki Aeroporia* [*File: Olympic Airways*], Athens: Stachy Press, 1992.

Cuba has increasingly become a worldwide tourist destination and as talk of relaxing the imposed embargo began to circulate at the turn of the century, air travel between the U.S. and Cuba was resurrected for a short period of time on a very limited scale. When and if the "all clear" signal is given to resume full air travel between the two countries, interested U.S. airlines will need to act methodically, yet strategically, in their quest to gain market dominance.

Technological advances affect every industry. Breakthroughs in computer technology or robotics can affect the speed and accuracy involved in the building and assembly of aircraft. American Airlines revolutionized the industry with their computer reservation system called Sabre. Sabre was completed in conjunction with IBM in the 1950s with it first mainframe going public in 1960. By 1964, Sabre was the largest private data processing system in the world.[6] Technology is a constant factor in every operation and any changes can significantly alter or make a business obsolete tomorrow.

Eclipse Jet and Friction Stir Welding[7]

An interesting example of technological changes and how it affects the airline industry can be seen in the case of Eclipse Jet's friction-stir welding technique and the evolution of the ultra-light business jet. Friction stir welding is highly automated and significantly faster than other structural joining processes. It enables a drastic reduction in aircraft assembly time and eliminates the need for thousands of rivets resulting in reduced assembly costs, better quality, and stronger, lighter joints. Friction stir welding will replace more than 60 percent of the rivets on major assemblies of the Eclipse 500 jet including the cabin, aft fuselage, wings, and engine mounts.

Finally, societal values and lifestyles can have a significant impact on the way a company conducts business. This fast-paced world has evolved into a commuter society. Today's commuter may have needs that necessitate travel from one coast to another. This increase in travel, due to dual careers and/or working a significant distance from home, affects airline and airport operations. Just as the healthy living trend affected the fast food industry in terms of food offerings and selections, lifestyle changes can cause a company to rethink its strategic direction or, at least, its product mix.

While these macroeconomic forces typically have considerable influence on a company's strategy, the greatest influence comes from the microeconomic environment – the industry and its competitors.

---

[6] See < http: //encyclopedia.thefreedictionary.com/SABRE%20reservation%20system>.
[7] See <http: //www.eclipseaviation.com/>.

The Micro External Environment

The immediate industry and competitive environment provides the playing field in which a company operates. This is known as the micro external environment and is depicted in Figure 4.2. Every industry has its own unique characteristics, its own unique competitive playing field, and its own growth potential. Industries, like people, have a life cycle ranging from infancy to maturity and death. Where an industry is in its life cycle can greatly affect the competitive environment. For example, in an emerging industry, competition is generally fierce as companies vie for market share. However, as an industry travels through adolescence it begins to experience shakeout, where only the strong survive. Industry life cycle will be examined in a later discussion on shakeouts; however, for the purpose of this discussion, it is important to realize that the economic characteristics are important factors in shaping the competitive playing field.

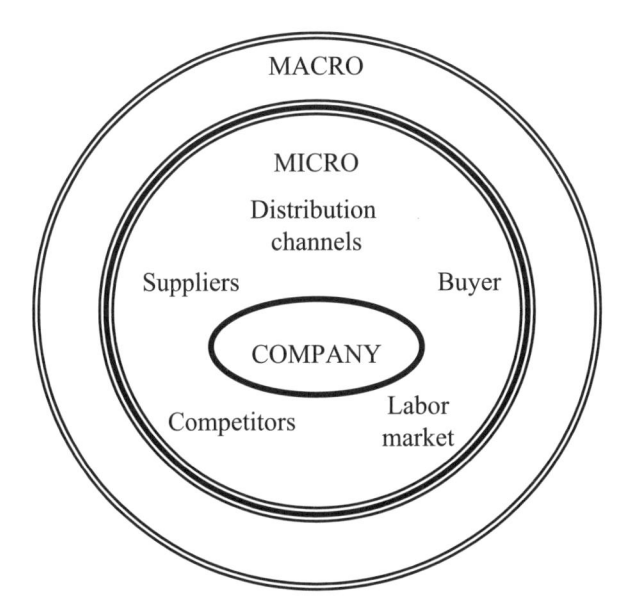

Figure 4.2 The Micro Environment

Some of the characteristics that affect the competitive playing field include the size and growth rate of the market, the depth and breadth of products in the market, the uniqueness of these products (or likeness as in commodity products), the existence or availability of distribution channels, the dependence and sensitivity toward technological changes, the make-up of the buyer and seller market and the effect of new entrants and economies of scale to the long-term success in the market. No two industries are identical and no two industries face identical competitive pressures. Some industries, like the airline industry, were introduced for the first time to the often cutthroat behavior of a competitive environment after many years of regulated complacency. Understanding the competitive pressures and how to exist within these conditions is paramount to industry survival.

Before moving forward in this discussion of analyzing the competitive environment, it is important to discuss the pressures that affect competition. Without a thorough understanding of these pressures and how they affect the businesses of an industry, it is nearly impossible to craft a sustainable competitive strategy.

Michael Porter's Five Competitive Forces

In 1980, Michael Porter published a framework for analyzing the competitive pressures in all industries.[8] While industries differ greatly in their composition, Porter argued that all industries face similar competitive forces. Porter's model of the Five Forces of Competition is presented in Figure 4.3.

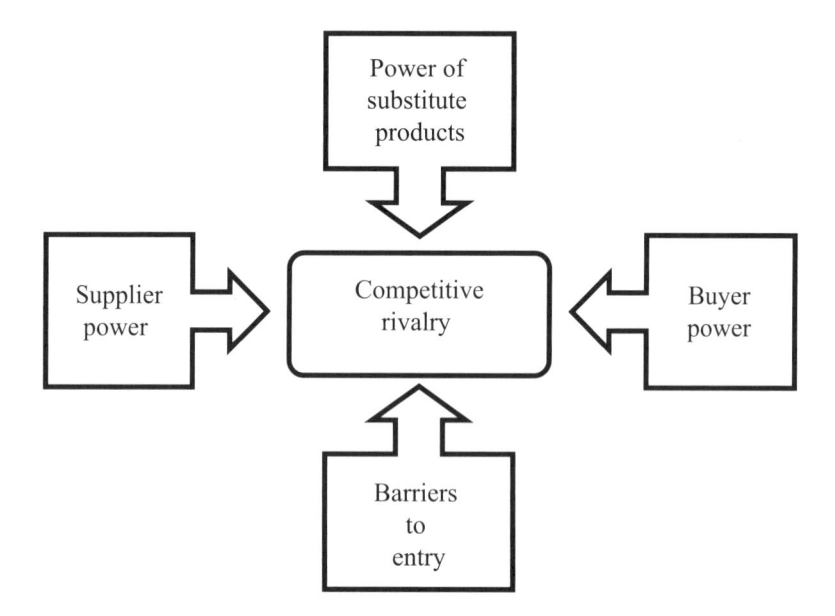

Figure 4.3 Porter's Five Forces of Competition

---

[8] Michael E. Porter, *Competitive Strategy: Techniques for Analyzing Industries and Competitors*, New York: Free Press, 1980, 108–122.

Porter's model provides a very simple yet methodical way to dissect an industry and determine the true pressures that affect the marketplace. The Five Forces model suggests that the competition in an industry is determined by five competitive forces:

1. Barriers to Entry.
2. The Power of the Supplier Market.
3. The Power of the Buyer Market.
4. Power of Substitute Products.
5. Rivalry among Industry Members.

Barriers to Entry

A barrier to entry into an industry occurs when it is difficult for a newcomer to enter into the market. New entrants in a market can change the complexion of the industry and almost always provide a challenge to the existing firms due to the newcomer's ability to carve out a portion of the market. However, sometimes it is difficult for a newcomer to gain a foothold in an existing market. There are many types of barriers that could provide a challenge to newcomers entering in a given market. Some of these are as follows.

Economies of scale Often the incumbent firms in an industry are large enough to have gained scale economies. Scale economies refer to the practice of receiving unit discounts for large quantity purchases. In other words, the more a company produces or ships at once, the lower the per unit cost becomes as large batches or shipping containers cost less than multiples of small batches or shipping containers. Therefore, a company that has gained scale economies pays less for materials than one that has not because the former purchases in bulk. A newcomer will need to enter the market on a large scale, which could be a risky move for a new firm, or they might have to accept the disadvantage of paying more for component parts, which could affect profitably. Either way, a natural reaction by the incumbent firms is to retaliate by slashing prices or increasing sales promotions. This move by the incumbent firm might likely be detrimental to the new firm who is trying to stay a float. Consequently, when scale economies exist in an industry, it serves to block new and emerging companies.

Inability to gain access to patents When technical capabilities are essential in an industry, patents held by the incumbent firms can serve as major barriers to any new company entering the marketplace.

Brand loyalty Customers are often attached to a particular brand. In the airline industry, for example, customers in given areas might be attached to a particular airline not only because of the service but also because the ability to accumulate frequent flier points. Establishing customer loyalty may be a long process for a newcomer. To overcome the pressures of brand loyalty a company must spend a great deal on advertising or provide some kind of extra service in order to lure the customer into switching brands, like provide cheaper fares to popular cities.

Regulations Governmental legislation and regulations can be a major barrier to new entrants. Prior to the deregulation of the U.S. airline industry, the regulatory environment kept new entrants out of the market. Today, the stringent safety and security regulations imposed on airlines after the disaster of 9-11 pose a new barrier to entrants because of the associated costs.

Access to distribution channels Incumbent firms can tie up the channels that provide a new firm access to customers. For a new entrant to succeed in this situation, the firm might have to bargain or buy its way into the market. This is another source of pressure on profits and can serve as a detriment to entry.

The learning curve Repetition leads to perfection. Often, the overall experience of providing the service or producing a product results in a lower cost to the firm. New entrants into the market will have the pressures of "know-how" to overcome.

Barriers to entry do not necessarily discourage a firm to enter a market, but they do provide interesting challenges to new and incumbent firms. When entry barriers are low, the threat of entry is stronger against incumbent firms. Another consideration for both the incumbent and new firms is whether or not the market can tolerate a new entrant based on present and future growth and profit projections. If there is sufficient room in the market, the risk to the new firm diminishes and the threat of entry is stronger for the incumbent. If the market is well saturated or has a little room for future growth, the threat of competition is a weaker competitive force.

The Power of the Supplier Market[9]

The supplier to an industry can wield a great deal of power. Consider the havoc a dual factory recall of O-540 crankshafts from Lycoming would cause to the aviation industry, since the crankshaft is widely used in aviation engines.[10] The supplier market is comprised of all the companies that provide component parts to a product. The power of the supplier market is closely related to the significance of the component part to the final product. If the component part is an essential part of the overall product, the supplier tends to have more control. However, if the component part is a

---

[9] Ibid.
[10] See <http://www.textronengine.litigation.com>.

commodity product, there usually is little to no competitive pressure, unless supplies become scare. In what way do suppliers have power over companies? When the component part is essential and there are few comparable products on the market, the suppliers have more leverage, for example, over the price and the terms and conditions of delivery.

On the other hand, suppliers have less power when the industry they are supplying is a primary customer. The loss of this customer could be disastrous to the supplier. The supplier in this case must be competitive in price, delivery schedules and the quality of the product supplied. However, when the component part being supplied is only a fraction of the supplier's business, the pressure on the supplier is lessened. At this point it is important to note that while previously quality was a major deciding factor in the selection of suppliers, today quality is a given. In other words, in order to even exist in the market, quality must be comparable. Businesses today will tolerate nothing less than the best. Consequently, firms today compete on pretty much the same playing field with regard to quality.

Another thing that suppliers must be concerned with is whether or not the industry members can make the product. In this situation, if industry members are dissatisfied with any aspect of the component part, they may choose to backwardly integrate. Backward integration means that the firm becomes its own supplier, making component parts. In some industries this threat is real. The more viable it is for a company to backwardly integrate, eliminating the need for the supplier firm, the greater the bargaining power industry members have over the suppliers.

To summarize the competitive pressures of the supplier market, industry firms have more to be concerned with when the suppliers have the bargaining power. Specifically, the power is with the supplier when there are few suppliers in the industry and when the component part is important to industry members. In this situation, the supplier is in command of the price charged, delivery schedules and all aspects of getting the product to the industry members.

The Power of the Buyer Market[11]

The buyer market is comprised of all the firms or individuals that purchase the product or service of the industry members. In the commercial airline industry, one segment of the buyer market is the passenger. The buyer is important and has significant purchasing power when the end user buys in large numbers. Why is the quantity important? Generally, when a buyer purchases in large quantities they are able to negotiate price concessions as well as other terms and conditions. For example, customers flying in large groups expect some kind of price concession from airlines. Travel agents are provided with volume discounts as well as corporations that have a loyalty agreement with certain airlines. Actually, several airlines have specific

---

[11] Porter, 108–122.

password protected areas designed on their websites to address the travel needs of their corporate clients. Other circumstances relating to the power of the buyer market include the following.

High switching costs If the buyer has the flexibility to choose among the industry members, the buyer gains significant bargaining power over the industry member. This situation would occur when there are many industry members and they are relatively identical in nature, such as two full-service commercial airlines servicing the same markets. However, if the product is different, it might be more difficult to switch brands because the learning curve is too high.

Small buyer market When the buyer market is small, the industry members have fewer choices to turn to if they lose a customer. Consequently, when there are few buyers, the buyer has the power over the industry members.

Buyer awareness Buyer awareness is another very important consideration for industry firms. Today, the Internet provides the buyer easy access to information about given products. Consequently, the more the buyer knows about a given product, the greater the bargaining power.

Buyer purchasing power Not all products are a necessity. If the price is not acceptable to the buyer, the buyer could choose not to purchase the product.

The competitive pressure of the buyer market is stronger when the buyer has leveraging power over the price, service, or any condition of the sale, including whether or not to purchase the product.

The Power of Substitute Products[12]

A substitute product is not equivalent to the product of a competitive firm. A substitute product is a product in one industry that can be substituted for a product in a different industry. For example, producers of glass bottles compete with producers of plastic bottles. If there is a substitute product available, the competitive pressure can increase. In the European markets, the train is a substitute for airline travel. Although the travel time is sometimes longer, if the airfares are high, frugal customers might select the train for their travel needs. As a result, a well positioned substitute can determine the ceiling on the price the end user is willing to pay; anything above this price results in the end user considering the substitute, because paying more will ultimately eat into the potential profits or, in the case of an individual, cut into spending money. Likewise, if the price of the substitute is less than that of the industry member's products, the

---

[12] Porter, 108–122.

competitive pressure lies in the fact that unless the member can find other means of cutting costs, profitability will be compromised.

The strength of the competitive pressure, therefore, depends on the availability of a substitute product, the ease in which end users can switch to the substitute product and whether the end user considers the substitute a viable alternative with regard to durability, quality and performance. The mere fact that a substitute exists will likely cause industry members to investigate alternatives because the bottom-line is ultimately the key to success. Thus, industry members are under additional pressure to prove to the end users that their product is superior. Alternatively, if by changing to the substitute product the end users will have to make some substantial changes to their organization, including changes in equipment or even assembly, then the likelihood of changing will be diminished. Overall, the competitive pressure of a substitute product is strong when it is readily available and when the product is reasonably priced.

Rivalry among Industry Firms[13]

Generally, the strongest of the competitive forces is the rivalry among industry members. The intensity and nature of the rivalry differs among industries. For example, in some industries price is the primary source of competition, while in others service, performance and brand image (to name a few) are key to the rivalry. The rivalry can be friendly or bitter, depending on the circumstances. No matter what the competitive arena, every company has determined a competitive position and a strategy for achieving that position. Success, then, is largely dependent on the reactions – offensive and defensive moves – of the competition. A good strategy is often one that catches rival firms off guard. But if a company has done its homework, it should be able to respond with a counter offensive move. Take, for instance, the airline industry: when a new carrier comes into a particular market the rival firm might react by offering specials or cutting prices on popular routes. The competitive battlefield may remain intense or it may taper off. In reality, the actions and reactions of the rival firms set the tone of the industry. Some industries are known for price cuts and advertising wars while others are more concerned with beating the competition in a more subtle manner, such as customer service or service after the sale.

Yet it seems that, no matter what the industry, rivalry is more intense when there are more competitors in the playing field. Often, as the competitors increase, the size and capabilities of the firms begin to equalize. At this point the competing firms are vying for market share and generally doing everything they can to dominate the market. However, when the firms are relatively equal, it is more difficult to become the dominant firm. This is when clever strategies and creative moves become more essential.

---

[13] Ibid.

Rivalry is also more intensive when the demand for the products in the industry is growing slowly. In every industry life cycle there comes a period of time when the marketplace is saturated and the demand for the products seems to taper off. As a result, a shakeout period occurs and only the strong firms survive. This creates an even stronger competitive playing field for the remaining firms. If one of the firms begins to lose market share, a natural reaction is often to act aggressively to try to bolster their market position. At this point one company in the marketplace might respond with the introduction of new products, price slashing and much more aggressive advertising. Other companies in the industry will usually follow suit and the competition intensifies again.

Rivalry is also more intense when it is easy for the customer to switch brands or, in the case of the airlines, switch carriers. If switching is easy, companies can run cost-cutting specials and steal away market share from a competitor. However, if switching costs are high, then rival companies tend to hold on to their market share and work harder to protect the interest of customers.

Another sure way to see increased rivalry in an industry occurs when the stakes are such that a successful strategic move can render high payoffs. The payoff can be considered high if it will take a competitor a long time to respond. For example, a company that introduces next-generation technology is often taking a risk because the technology has not likely been perfected and is subject to problems. At the same time, however, rival firms are going to have more difficulty responding to next-generation technology and, therefore, their competitive move occurs more slowly. Meanwhile, the initiator of the strategy to introduce next-generation technology is able to reap the benefits of its strategic move without retaliation from the rival firms.

While there are numerous scenarios that exist where rivalry may appear strong or weak, one thing remains certain: if rivalry intensifies there is always the concern that company profits will be affected. Consequently, rivalry among competing firms is one of the strongest and most important forces encountered in the competitive arena.

Industry Characteristics

Every industry has a unique set of characteristics. Having a thorough knowledge of these characteristics and understanding what makes the industry tick is essential to marketplace success. For example, while unemployment might be high nationwide and, consequently, firms can be selective in hiring, in a given industry one might find that qualified individuals are quite rare. In a situation like this the lack of pressure from the macro environment is overshadowed by the pressure to find well-qualified individuals trained in a specific industry or field. Consequently, the key to good environmental analysis is to be aware that every industry is different. As will be seen in Chapter 8, some corporations are made up of businesses in several different industries. To best assess the direction the corporation should take in the future, each of the industries in which the corporation's companies compete must be considered.

Understanding Opportunities and Threats in the Industry

An important point in studying the external environment is to understand opportunities and threats. An opportunity is something that a company can take advantage of that might help in gaining a competitive advantage.

Table 4.1 Opportunities for the Aviation Industry Post-September 11

Opportunities
The concept of fractional jet ownership has taken off, making it easier for even small businesses to have access to business jets to use for company travel.
There is growth potential in the ultra-light business jet market, such as the Cessna Mustang, Eclipse 500, Diamond Jet and Adam Aircraft.[14] This can create a new business niche for air taxi services in the future.
Several business jet types, such as the Falcon 7X and Cessna Citation X, can fly very high and near supersonic speeds, thus shaving off travel times for long-haul flights.
Airline Alliances of various forms have reached maturity and range, from codesharing to strategic alliances and virtual global networks.
Last-minute deal internet clearing houses (e.g. Orbitz, Priceline and Hotwire) provide more options to travelers.
Low-cost airlines worldwide (e.g. Virgin Blue, Southwest, Ryanair, AirAsia and Air Arabia) provide more flexibility and lower prices to travelers.
Liberalization of air traffic services worldwide has the potential to further stimulate the air travel business.

For example, SkyEurope Airlines, a budget airline out of Slovakia, saw an opportunity at the airport in Bratislava, Slovakia. The airport had a capacity of 2 million passengers but was only serving 300,000. The excess capacity, coupled with

---

[14] See <http://www.avweb.com/bizav/9_04/complete182466-1.html>.

the fact that there was little competition and high demand for air travel, attracted a former consultant to start up SkyEurope. In addition, the lower airport and labor costs in Slovakia, as compared to neighboring Austria, were yet another opportunity for SkyEurope. Why? Since the cost of doing business in Slovakia was lower, SkyEurope could compete against the high costs at the airport in Vienna, which was about 30 miles away. For the difference in price, the customers were willing to drive the short distance. In effect, Bratislava became an alternate airport to the Vienna airport.[15]

A threat is a condition in the external environment that may constrain a company in trying to gain a competitive advantage. Let us look at the SkyEurope example, again. Prior to SkyEurope, the Vienna airport was attracting many customers from the southern part of Slovakia because it was only a short drive from Bratislava to Vienna. With the advent of SkyEurope, the Bratislava airport became a threat for the Vienna airport. Why? First, the Vienna airport is one of the most expensive airports in Europe. Specifically, flying out of Bratislava saves the passenger about $17 USD per ticket in airport taxes. With salaries considerably lower in Slovakia, SkyEurope's labor costs were lower, thus allowing the airline to pass on this savings to its customers in the form of lower airfares, which were much lower than the carriers using the Vienna airport. As a result, the Vienna airport lost both the Slovakian passengers and also some of the Austrian travelers. If SkyEurope begins flying to an even greater number of destinations, this could be a major constraint for the Vienna airport and the carriers serving that airport.

Table 4.2 Threats to the Aviation Industry Post-September 11

Threats
Acts of air piracy and terrorism can prevent growth in the airline business.
Financially solid, low-cost carriers have been challenging legacy carriers in several business segments.
Corporate travel budgets have shrunk after 9-11, producing adverse effects on business travel.
Advances in telecommunications have reduced the need for face-to-face business meetings.
People demand cheaper travel due to increased competition, putting pressure on razor thin profit margins especially on legacy carriers.

---

[15] See <http://www.skyeurope.com/start.php?lang=en&agencyId=>.

Threats and opportunities are the external components that make up what is known as SWOT analysis. SWOT stands for Strengths, Weaknesses, Opportunities and Threats. We will discuss strengths and weaknesses in Chapter 5. The SWOT analysis provides a clear view of the company's internal resources, both capabilities and deficiencies, as well as its external market opportunities and constraints. The premise of SWOT is that the goal of the strategy-making process is to produce a good fit between the company's internal capabilities and the environmental conditions. Figure 4.1 depicts opportunities and threats in the aviation industry post-September 11.

Driving Forces

It is seldom that an industry remains stagnant and immune to trends and developments; in fact, it is doubtful that any industry can ever be immune to such changes. At the very least, all industries go through a life cycle similar to that of human organisms. Industries are born, go through a period of growth or adolescence, mature, and eventually wither and die. Obviously, the changes in the growth cycle of the industry alone affect change in the industry. However, there are many more reasons for change than just the industry life cycle; there are also driving forces.

What is a Driving Force?

A driving force is something in the competitive arena or in the industry itself that creates some kind of momentum or pressure for change. What might be considered a driving force? Let us consider a few of these.

- *Changes in Technology* As technological innovations are introduced they can be a driving force for change. In some cases, companies have no other option but to change with the new technology. Often, not responding to the changes in technology can put a company behind the learning or experience curve, or it can place the company behind, competitively, because the new technology could be one that will lower the cost of doing business. These changes can result in major changes in the capital requirements in an industry as well as the expertise of the staff.
- *Increased Globalization* As globalization increases, it changes the industry's landscape. Both domestic and international companies might launch efforts to gain dominance. In order to compete, this might affect pricing strategies, or it may improve the ability to locate niche markets – markets where the larger, global competitor cannot afford to enter.
- *Entry or Exit of Firms* As previously noted, when we talked about Porter's Five Forces of Competition, the entry of new firms can change the competitive playing field. As new firms enter the industry, the existing firms must find a means of

coping with the new competitor. Sometimes, the new firms enter on such a small scale that it could possibly not affect the business of the major players in the industry. However, as was the case when Southwest first entered the market, the new competitor may prove to be a thorn in the side of the existing companies because it intrudes in the market share of the existing firms. The exit of a firm can also change the playing field by providing new territory for competitive ventures. When a void of this nature opens, companies must determine, first, whether it is worth their while to try to absorb some of the void and, second, how they will try to compete.
- *Demographic Changes*   New ways to use a product or changes in the people actually using the service or product can be another driving change in the industry. This can mean that younger people are now using the product or service, or it could be that the change in demographics, as a result of longer life expectancies, is a driving force for change.

Understanding the Competition

An important part of understanding the environment is to understand the competition, and a good way to know the competition is to try to categorize it in terms of the services that they offer, market that they serve, prices that they charge, costs that they incur for their services and technological approaches that they take. In essence, a company needs to know everything there is to know about its competition. By looking at the competition in terms of categories, it helps to see who exactly the competition is. These categories are often referred to as strategic groups. An industry has only one strategic group if every company in the industry is trying to attain the same strategies and market positions. However, this is uncommon.

To determine the strategic groups of the industry, first it is important to identify all of the different characteristics that separate the firms in the industry. Some of the identifying characteristics might be price range, geographic coverage, depth of the product line and the degree of services offered. After identifying the characteristics, it is often helpful to plot these characteristics according to two anchor points. For example, on one axis high priced and low priced might be measured, and, on the other axis, wide geographic coverage and regional coverage. Then, the companies that seem to fit together into groups are assembled. If the companies seem to be relatively close in terms of their characteristics, it can be deduced that the competitive playing field is intense. Figure 4.4 depicts several carriers in the United States along the lines of geographical scope and price. The legacy carriers score higher in terms of price and have extensive domestic and international networks, whereas low-cost-carriers have purely domestic or limited international networks (e.g. JetBlue) and lower average ticket prices. Regional carriers are typically limited in geographic scope, albeit a few exceptions exist, and they typically have higher price levels that reflect higher

CASM (cost per available seat mile) structures.[16] Figure 4.5 depicts several carriers throughout the world, just like Figure 4.4 did for the U.S., along the lines of price levels and geographical scope. In Figure 4.5 we see that several airlines fall in a new category altogether, called purely international. These airlines only fly international services either because they are the designated international airline of their countries (e.g. Air China and Air India) or there is no domestic network in their country (e.g. Singapore Airlines).

SCOPE

PRICE	Regional	Purely domestic	Domestic/int'l	International
Low		Southwest Air Tran	JetBlue	
		America West Frontier		
	Cape Air American Eagle Sky West Mesa		Comair ASA	Northwest United American Delta US Airways Continental
High		Midwest Express		

Figure 4.4   Examples of Several U.S. Carriers in Terms of Geographic Scope and Price

One of the next things on which to concentrate is what competitive moves the rival firms are likely to make. Companies that do not pay attention to the competitive moves of rival firms might as well be operating in a bubble. No company can outsmart its competition unless it is savvy and astute. Like all competitive sports, a company wants to be on the offense, anticipating the next step rather than becoming wrapped

---

[16] This notation is somewhat problematic for several of the airlines listed, for example, ASA, which has an international network that reaches Canada and serves several city pairs that would not be considered regional, such as Atlanta to Montreal.

up in defending a position after the other team has made great strides invading their territory.

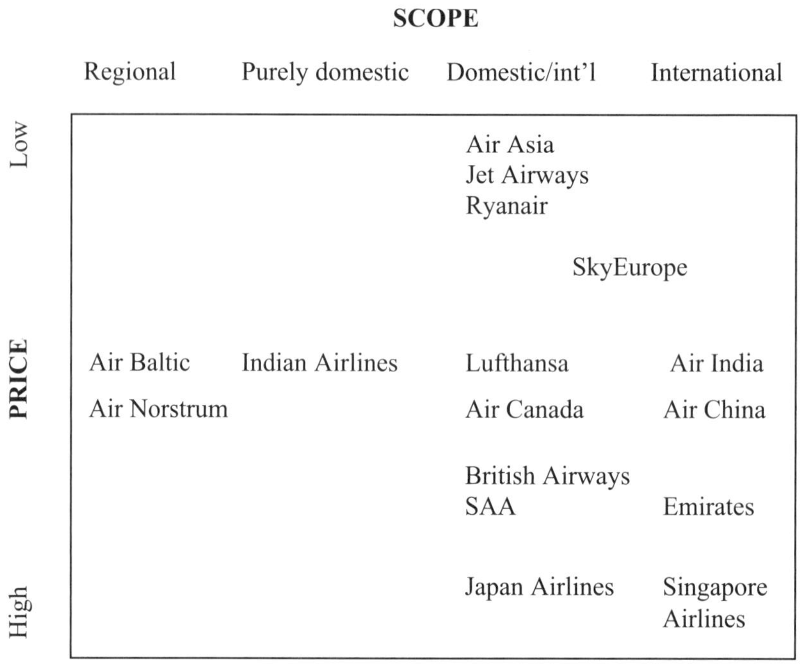

SCOPE

	Regional	Purely domestic	Domestic/int'l	International
Low			Air Asia Jet Airways Ryanair	
			SkyEurope	
PRICE	Air Baltic Air Norstrum	Indian Airlines	Lufthansa Air Canada	Air India Air China
			British Airways SAA	Emirates
High			Japan Airlines	Singapore Airlines

Figure 4.5   Examples of Several Non-U.S. Carriers in Terms of Geographic Scope and Price

How does one obtain information about rival firms? Study them: closely examine their geographic coverage, market share, competitive positioning and willingness to take risks. A company must know everything possible about their competition. They must study the annual reports of the company and their 10-K filings and keep a watchful eye on their webpage and special promotions. It is to the company's advantage to gather information constantly. While it is seemingly difficult to determine the competitive moves of rival firms, the more a company monitors the moves of these firms, the better their chance of never being blindsided by an unexpected move. What is the wrong way to gather information about rival firms? By breaking the law, as it seems WestJet allegedly did with Air Canada in 2004, when it used unauthorized access to Air Canada's intranet (by a former employee of Air Canada employed by WestJet) to monitor Air Canada's actual and forecasted passenger loads and other information.

Key Success Factors

What is a key success factor? Key success factors are those factors that positively affect a company's chance to prosper in the marketplace. In other words, they are the characteristics shared by truly successful companies. If a company is successful in a given industry, like the airline industry, then this company must excel in certain things in order to be both financially and competitively successful. It can be product attitudes, competitive capabilities, or even market achievements.

It is important to determine the key success factors in an industry. Managers have to be aware of what it takes to be successful competitively and what factors are unnecessary to success. Key success factors vary from industry to industry and can even be affected by time, as driving forces in the industry can change what is important at any given time. Key success factors can be related to technological advances, marketing skills, cost advantages, customer service advantages, information systems capabilities, patent protections, or brand identity. As a manager, it is prudent to be able to diagnosis, accurately, what factors are most important to success. Misunderstanding key success factors can result in misdirected strategies and failure in the marketplace. Key success factors truly are major opportunities for competitive advantage. At the very least, as a top manager it is important not only to know the key success factors but also to be able to evaluate his or her company against the key success factors in the industry. This allows one to determine where a company's major areas of weaknesses are and what must be done to change that company's position in the competitive arena.

Another exercise a manager might engage in is to use the key success factors to compare his or her company against their major competitor. Often referred to as a competitive matrix, it is a very subjective method of assessing a company not only against the major factors for success in the industry but also against the competition. To complete a competitive matrix the key success factors are listed in a column on the left-hand side. Next, based on a total score of 100, a weight is assigned to each of these factors relative to one another. For example, if market share is the most important, it might receive a weighting of 0.40, whereas customer service might receive a weighting of 0.20. Then all the companies are rated according to a five-point scale, where five means the company is very effective (does an excellent job) and one means the company is ineffective (does a poor job). After rating all of the companies against all of the factors, the weight given to that factor is multiplied by the rating assigned to the company. For example, if Airline 1 received a score of two for market share and the market share was weighted .25, then the weighted score for Airline 1 for market share would be .80. Finally, all of the weighted scores are added together for each airline, which produces the total weighted score. This score will help to determine not only where the company stands in terms of the key success factors but also where it stands with regard to its competitors. Table 4.3 provides an example of a competitive matrix for the airline industry.

Table 4.3 Competitive Matrix for the Airline Industry

Key Success Factors	Weights	US Airways	Weighted Score	American Airlines	Weighted Score	Delta AirLines	Weighted Score
Market Share	0.25	2	0.5	5	1.25	4	1
Customer Service	0.15	2	0.3	3	0.45	4	0.6
Price	0.35	5	1.75	3	1.05	3	1.05
Frequent Flyer Program	0.15	2	0.3	5	0.75	4	0.6
In-flight Passenger Comfort	0.1	2	0.2	5	0.5	3	0.3
Total Score			3.05		4.00		3.55

Obviously, the biggest drawback of the competitive matrix is that the weightings and ratings are subjective. While this is not a scientific approach to analyzing the competition, it does provide one more tool for assessing industry competitors. The more angles that are explained, the better the final assessment.

The Process of Environmental Analysis

There are four major components of environmental analysis: environmental scanning, monitoring, forecasting and assessing.

Environmental Scanning

Environmental scanning looks at all possible angles of the environment. In scanning the environment it is important to be alert to any changes that are taking place in the environment as well as any early signs of potential change. Obviously, the information is not going to be crystal clear. Information will likely be rather ambiguous and incomplete, but this can still give an indication of what is on the horizon. Environmental scanning is of the utmost importance in industries that are always changing, like the airline industry. For instance, a possible increase in minimum wage should be noticed by industry scanners, because this can have an effect on the entire wage system for companies that employ any number of individuals at this level. Where would the increase in dollars for wages come from? Either straight out of profits or in fare increases – and these are issues that must be planned for in advance.

Monitoring

Monitoring means to track continually the issues that are noticed in environmental scanning in order to confirm or disprove things that were identified in the scanning process. Monitoring is somewhat less broad than scanning. In this case, a manager tries to assimilate all of the possible information available around the issue identified in the scanning process. For example, there might be talk of a new airline coming into a major service area. While the "talk" is just speculation in the scanning phase, in the monitoring phase a manager will try to determine the legitimacy and timeframe of the rumor.

Forecasting Environmental Change

Forecasting environmental change is the process of trying to predict the future trends and events the organization is monitoring. This is done through a process of trend identification and extrapolation. In forecasting events, the manager tries to answer the following question: If these trends continue, either at an accelerated trend or at the present trend, what will the issues be in the future?

Strategists try to determine feasible projections of what could possibly occur as a result of what has been found through the scanning and monitoring process, and then determine how quickly the change could actually have an affect on the industry and the company.

Assessing Environmental Change

Trend identification and extrapolation is a means of plotting out the information gathered in the environmental analysis and then projecting future occurrences. In looking at these trends, it is the responsibility of the manager to determine a series of alternative projections. The process is more exact and easier to accomplish if good financial or statistical data exists. How often does this occur? Not very often! Nothing in the environment comes packaged neatly. Most of the environmental issues are poorly defined and unclear. Therefore, it is up to the management team to come to a consensus about what is going on in the environment today and speculate on the future.

To do a truly good job at trend identification and extrapolation, management needs to look at the different categories of environmental issues, issues of economy, politics, demographics and legislation, just to name a few. Next, it is important to determine whether these factors are opportunities or constraints. In an effort to try to work out the best consensus, managers in many organizations plot these trends on a graph

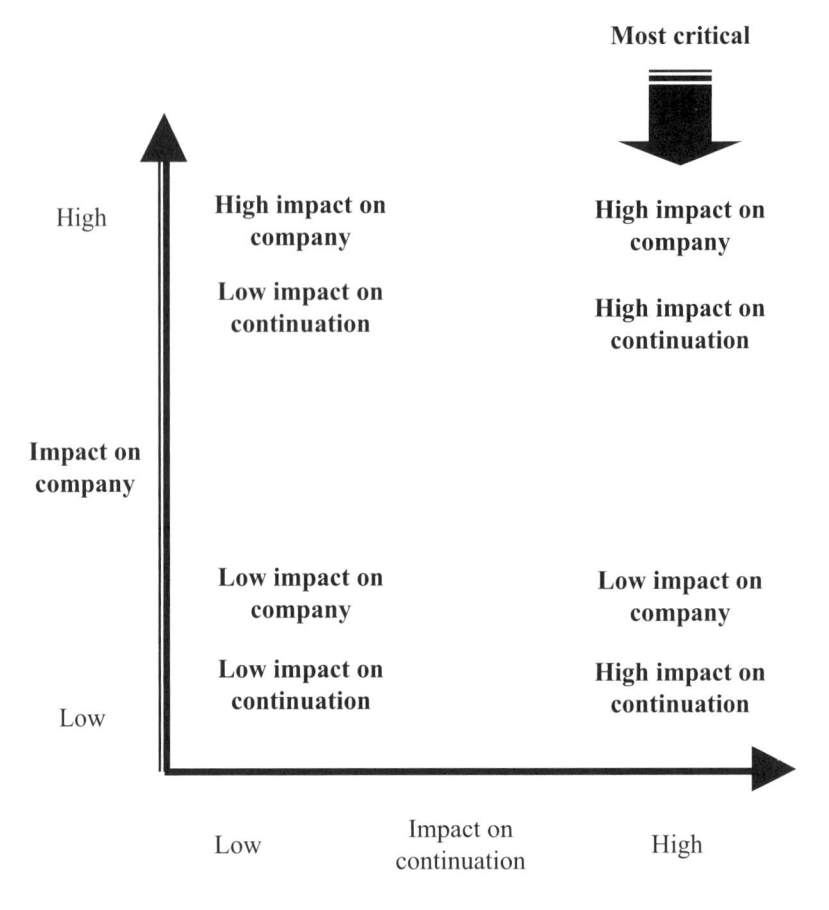

Figure 4.6   Assessing Environmental Change through Plotting Trends

where the y-axis represents the impact on the organization and the x-axis represents the probability of the present trend continuing into the future. By plotting the trends on such a graph, as illustrated in Figure 4.6, it is much easier for the management team both to see the issues and get a grasp of what they are.[17]

Using Experts to Help in Environmental Assessment

Many companies use experts to help them in their environmental assessment. The experts are outside sources, who can extend the thinking and broaden the perspective

---

[17]   Peter Ginter, Linda Swayne and W. Jack Duncan, *Strategic Management of Health Care Organizations*, Malden, MA: Blackwell, 1995, 95.

of the management team. People who have studied the airline industry or have served in different aspects of the air traffic industry are good choices to broaden and neutralize the perspectives of the management team.

The Delphi Technique

The Delphi Technique is a common tool used in environmental assessment. It is used to study, in more detail, a specific topic, current trend, or emerging event and is conducted in a series of rounds. In the first round, a panel of experts are asked their opinion on a specific topic, for instance the possibility of a terrorist attack happening as a result of current security procedures at a specific airport. The opinions of all the participants are then summarized and sent back to the participants for them to develop new thoughts and judgments on the topic. This may go on for several rounds in order to revamp ideas and summarize thoughts. In the end, the group has come up with a combined opinion on the topic.[18]

Revised forms of the Delphi Technique have arisen over the years. Some experts believe that the Delphi Technique could be best accomplished without getting all of the experts in the same room. S.C. Jain revised the Delphi Technique to include the following: [19]

- Identification of experts in the field.
- Individual examination by experts of a brief, based on a detailed analysis, on the issues in the industry.
- Detailed, structured interview with the experts on the issues addressed in the brief.

What makes this method a bit more appealing is the fact that the experts do not have to devote as much time to the process and, therefore, one might have an easier time getting them to cooperate. Of course, another positive aspect to this method is that all the egos are not in one room competing for center stage! However, from the standpoint of the company, Jain's method is much more time-consuming. The researcher has to do a great deal of work in order to prepare a detailed brief, and the structured interviews can consume a great deal of time. The result of this process may be a wide array of converse opinions, or it might be a succinct document from which to base future decisions.

---

[18] James L. Webster, William E. Reif, and Jeffery S. Backer, "The Manager's Guide to Strategic Planning Tools and Techniques," *Planning Review*, 17, no. 6, 1998: 4–13.

[19] S.C. Jain, "Environmental Scanning in U.S. Corporations," *Long Range Planning*, 17, 1984: 66–75.

Organizational Brainstorming

Sometimes one of the best ways of taking the information gathered in the environmental analysis and making some sense out of it is through organizational brainstorming. Organizational brainstorming entails identifying a group of the influential and/or knowledgeable individuals in an organization (maybe department heads) and bringing them together for the purpose of trying to understand the issues, assess the impact of these issues on the organization, and generate some strategic alternatives. In a well-structured brainstorming session, everyone has an opportunity to present an idea and to clarify an idea. The ideas are recorded, and no one is permitted to evaluate or criticize any idea. Basically, in a brainstorming session, anything goes. Nothing is considered outrageous, impossible, or risky. The point is to foster thinking and stimulate creativity so that new and fresh ideas emerge, which might then give way to new approaches. Brainstorming is not always the answer to new strategic directions; however, if individuals are willing to follow their creative impulses, it sometimes can help the organization find a new strategic niche.[20]

Tools to Address Environmental Uncertainty

Sometimes, when uncertainty is high, managers need additional tools to assess the best future. These tools are used to complement rather than replace the other components of the strategic process. One such tool is scenario planning.

Scenario Planning

Scenario planning is a systematic approach to imagining scenarios in the future. Each scenario tells a story of how the future might unfold and how critical elements might interact. It helps to develop a large range of possibilities, and the scenarios actually help identify the potential risks and alternatives. Scenario planning is most useful when the levels of uncertainty are high for several critically important assumptions. How does one go about using scenario planning? First, it is important to identify the fundamental questions that need to be answered. Then, identify the greatest areas of uncertainty that are impacting these fundamental questions. Next, develop the best and worst-case scenarios. It is important that these are the plausible extremes for the given situation and the given level of uncertainty.

After the scenarios are developed, elaborate on the scenarios and write descriptions of what could happen. One needs to be as detailed as possible. After the scenarios

---

[20] J. Stewart Black and Lyman W. Porter, *Management: Meeting New Challenges*, Upper Saddle River, NJ: Prentice Hall, 2000, 256.

are well thought out, determine all potential outcomes. These scenarios help paint a better picture of what the future could hold.

When to Use Scenario Planning:

When the company has experienced too many costly surprises in the past.

When the organization does not usually generate new opportunities.

In order to establish a common framework in thinking about the future.

The Limitations of Environmental Analysis

Environmental analysis is essential to effective strategic management and strategy formulation, but there are no guarantees that a good assessment of the environment will lead to success. Obviously, one of the biggest problems with environmental analysis is that it is conducted by humans, and humans and their projections are not perfect! In other words, managers cannot see everything. Similarly, without a crystal ball, environmental analysis cannot perfectly foretell the future. Another problem is that sometimes there is very important information necessary to the analysis, but there is no way of obtaining the information. Much of the information a company needs to predict the future is based on the moves of others, and this information, for the most part, is proprietary. Still another limitation to environmental analysis is red tape. Some organizations are so bureaucratic that they lack the ability to respond quickly enough to the issues and events detected in the environmental analysis. To respond to environmental issues, particularly in ever-changing or turbulent industries, companies must learn to cut quickly through the red tape. Finally, some top managers do not want to see what the environmental analysis is showing them. Sometimes managers are so set in their beliefs that no amount of evidence is going to get them to see the future clearly. Often, these managers used their own filtering mechanisms to block out what is obvious, or they discount the information on the basis of lack of full information. Whatever the case, it is generally managers like these who create the greatest obstacles to good strategic decision-making. Even the best and most comprehensive analysis is not going to reveal everything taking place in the environment, but managers must remain open to the wealth of information that can be obtained from a well-conducted environmental analysis.

Case Illustration: The Denver International Airport: An Environmental Debacle[21,22]

In November 1989, ground was broken to build the Denver International Airport. When the airport finally opened in late February 1995, it was 16 months behind schedule and close to $2 billion over budget. However, one of the most important issues that surrounded the construction of the Denver International Airport was the fact that economic projections prior to construction showed little support for the airport.

Building the new airport was initially justified in order to solve a capacity problem. Denver's economy grew in the early 1980s due to booms in the oil, real estate, and tourism segments of the economy. It seemed at the time that the old Stapleton Airport was limiting the attractiveness of Denver and its surrounding areas to businesses that wanted to relocate to the area. As the situation existed in the early 1980s, it seemed that delays at the airport were common and that a new, expanded airport was necessary. The airport also became a major issue in the mayoral election in 1983.

Ultimately, the capacity problem never materialized. Since the airport expansion was a significant issue in the election of the new mayor, he remained strong in his support of the new airport. By 1986, after the political commitment by the new mayor to expand the airport was made, but prior to the commitment of federal money to the project, it became clear that the forecasts for activity at Denver's current airport, Stapleton airport, were overstated. Specifically, the forecasts had not taken into account the impact of post-airline deregulation, the impact of the acquisition of Frontier Airlines by Texas Air and the bankruptcy of Continental Airlines. Instead of rising, Stapleton's share of the total U.S. domestic market fell by over 4 percent per year, and this decrease was over and above the decrease experienced nationwide at airports. At the same time, the economy in the Denver area took a sharp downturn. All of these factors combined indicated that there was little need for the new airport.

Another problem that arose was the fact that those in charge of building the airport did not get commitments from the major airlines prior to building the airport. Only after construction had begun were commitments made and, as a result, much of what was already constructed had to be changed in order to meet the demands of the client airlines.

What were the problems? First, there was no justification for the new airport. Those involved allowed politics to get in the way of good solid business decisions. All of the environmental factors pointed to the fact that capacity would not be a problem, yet because of the political agenda of the mayor the strategy of building an airport was implemented. Another problem was the fact that the airport authority did not

---

[21] Lynda M. Applegate et al., "BAE Automated Systems (A): Denver International Airport Baggage System," *Harvard Business School Case*, 1996: 1–15.
[22] Applegate, 1–3.

pay attention to its customers. Without getting commitments from customers at the beginning of the project, a great deal of money was wasted in reconstruction.

Chapter 5

The Internal Environment

> We know we are only as good as our last arrival. And if you're going to have an airline, you better hit the center line on the runway every time.[1]
>
> David Barger

In a highly competitive environment like the aviation industry, the players need to develop true capabilities in core areas, de-emphasize the less important or less profitable areas and fix what is currently broken. Service industries of all types are finding themselves under increasing pressures to cut costs while at the same time provide even greater customer service.[2] This is particularly true in the airline industry. The airline that successfully navigates the changes necessary to compete in this new arena will likely establish industry leadership and will have the opportunity to stay ahead of the competition. The airline that can set a benchmark for customer service and efficiency will survive even during the most turbulent times. How does an airline get into the position necessary to address these needs?

Chapter 5 focuses on the internal environment. The internal environment encompasses all of the relevant forces within the company's control. It is possible to look at all aspects of the internal environment individually and see a viable, healthy organization. However, when these same aspects are viewed in tandem, or when one examines how all of these aspects fit together, a totally different picture may emerge. It is how well the parts fit together that leads to a more efficient, well-run organization. A change in one aspect of the organization could force a company to change its business model or strategic direction based solely on a single part of the business.

In corporate planning the primary focus is on those factors in the internal environment that affect the company as a whole. Understanding the internal environment means understanding to the fullest extent possible the company's resource capabilities, relative cost position, financial picture, and competitive strengths as compared to its rival companies in the industry. To begin with, six questions should be addressed by a company:

---

[1] David Barger quoted in Bruce Schoenfeld, "We're Fed Up!," *Cigar Aficionado*, July/August 2002, 59.

[2] Sharon L. Oswald and William R. Boulton, "Obtaining Industry Control: The Case of the Pharmaceutical Distribution Industry," *California Management Review*, 38, 1993: 138–162.

- How can our core competencies drive success in our marketplace?
- What are our key customer groups and how do we secure them?
- How do we differentiate ourselves from our competitors?
- If our core competencies are replicable, what is our risk?
- What actions can be taken to restructure the business around core competencies?
- Which capabilities require development and which should be divested?

While the answers to these questions must take into account both the internal and external pressures, let us first focus on the issues within the realm of the company's control.

Value Chain Analysis

As has been suggested throughout this book, in order for a company to compete effectively in an industry, there not only needs to be a thorough understanding of the external environment but also a good understanding of the individual company's capabilities. In 1985, Michael Porter introduced the concept of Value Chain and Strategic Cost Analysis, and this tool is still used today to examine a company's environment.[3] Figure 5.1 depicts the linked set of activities that make up the value chain.

Value chain is a tool used to help separate the activities, functions and business processes necessary in running a business. The value chain approach helps to analyze processes involved in directly providing the service or producing the product, which are known as primary activities, as well as the processes involved in supporting the service or production, which are known as secondary activities. The value chain of a business should be seamless among every element of the chain. This means seamless within the business.

Each of the activities in the value chain is an important business function. In order for these functions to be "value-creating," the firm must determine ways that the functions can be conducted more efficiently, less costly and/or with greater quality. The activities are defined in the box 'Defining the Value Chain'.

While all of the primary activities listed above probably make logical sense, one might wonder why profit margin was included since it is a given that most companies expect to make a profit and pass these extra savings on to the customer. The aim of value chain is to create enough value that it is far and above that which is passed on to the consumer. Although profit margin is not considered to a primary activity, it is expected to be the outcome of the value-creating activities.

---

[3] Michael E. Porter, *Competitive Advantage: Creating and Sustaining Superior Performance*, New York: Free Press, 1985, 33–61.

Primary activities

Inbound logistics → Operations → Distribution and outbound logistics → Sales and marketing → Service → Profit margins

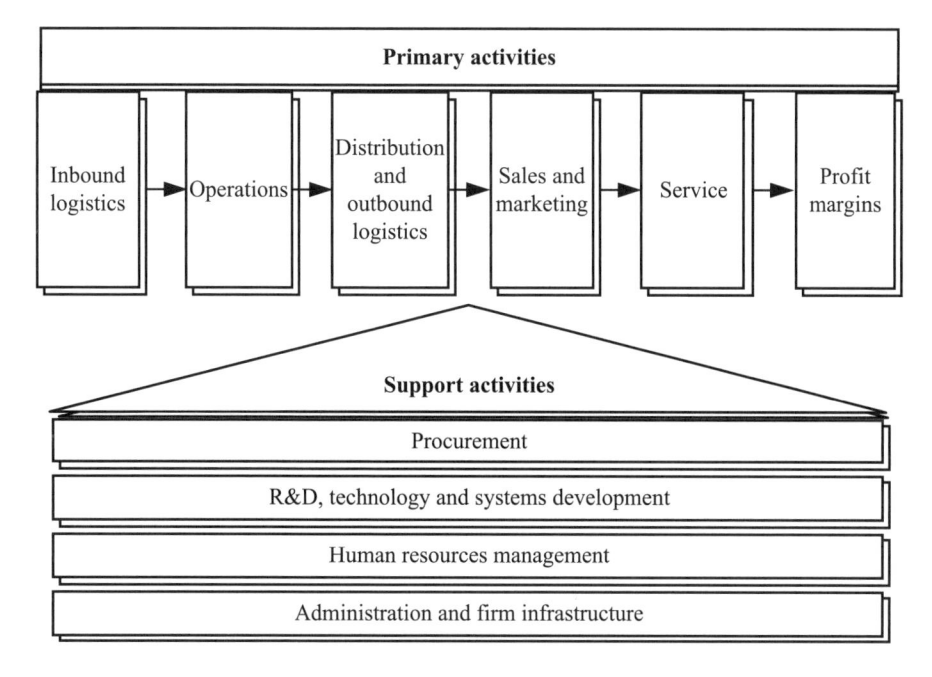

Support activities

Procurement

R&D, technology and systems development

Human resources management

Administration and firm infrastructure

Figure 5.1   Value Chain[4]

How Does One use the Value Chain?

How does a company use the value chain to strengthen its organization and capitalize on that value? It is likely that every company in the industry has similarly configured value chains; therefore, the key to using the value chain is to study how different the configurations really are. Why? It is within these differences that a competitive advantage can be found. It is the total package that the firm creates that provides the value, and it is the customer who decides whether or not the value created is the best value. For example, the firm should strive to produce the product or service at the lowest possible cost, thus, allowing the company to potentially undercut the competition. Or the firm might look to capitalize on certain attributes that meet the customer needs better than competing firms. Value chain analysis helps to locate areas in the company's processes where cost reductions or improvements can be made to the end value. The end objective is to optimize the difference between value and cost. It is the linkages between the different aspects of the value chain that can make the competitive advantage difference.

---

[4]   Ibid.

Defining the Value Chain

Inbound Logistics: These are all the activities involved in product inputs, such as receiving, storing, handling, and disseminating all the material necessary for production or service.

Operations: All of the activities involved in converting the inbound logistics into a product or service.

Distribution and Outbound Logistics: The activities involved with delivering the final product or service to the final customer. In a production operation, this includes warehousing, materials handling and order entry.

Sales and Marketing: The means by which products or services are made available to the public. This includes methods that convince the customer to purchase the product or services.

Service: The activities involved in making the product special. Those value-creating activities that set the product or service apart from the competition.

Support Activities[5]

Procurement: All of the activities necessary to purchase the inputs. This includes raw materials, supplies, office materials and bricks and mortar.

R&D and Technology and Systems Development: All activities to improve the product or the process involved in producing the product or providing the service. This could include product design and service procedures, to name a few.

Human Resource Management: The activities involved in recruiting, hiring, compensating, training, promoting and terminating employees.

Administration and Infrastructure: These are all the management activities including planning, legal counsel, budgetary issues, general business office activities, legal support and other related activities.

---

[5] Ibid.

It is important to realize that the value chain, while seemingly internal, does not exist solely within the boundaries of the company. While there are linkages within the firm, there are also important linkages outside the firm, and these linkages connect the firm with both suppliers and customers. Customer focus groups, for example, have helped companies gain a better understanding of the needs and desires of their customers, providing a very important bridge between the company and the end user. Certainly, most firms realize the need to work closely with manufacturers and suppliers, because this is essential to production. Yet not all firms stay close to the end users; instead, sometimes they remain isolated from the marketplace.[6] It is the customer's perception of value that is most important, ask any executive in business today.

Figure 5.2 depicts a generic passenger airline business processes model. Some of the core processes are customer focused (shaded areas) and some are non-customer focused (non-shaded areas). The passenger airline value chain depicts those processes directly involved in providing the air transportation service, which are known as primary activities, as well as the processes involved in supporting the service, which are known as secondary activities. For the airline value chain to be effective, the processes or activities involved in and making up the value chain, both primary and secondary activities, should be seamless among every element of the chain. In other words, the value chain should be seamless within the airline. Later in this chapter we will show how airlines can use an integrated software approach to help them achieve a seamless construction of their activities.

What pleases airline passengers and helps airlines score high on perceived value by passengers and, in turn, adds to the bottom line of the company? Is it more space for coach passengers' legs or more in-flight personal entertainment systems? Most passengers expect in-flight entertainment systems and some U.S. carriers believe this should be expanded beyond their international fleets. While passengers enjoy more space, American and United Airlines have found that more-spacious coach seating generally does not pay off financially. The premium an airline might charge for more-generous coach seating is exceeded by the revenue lost from seats removed. On the other hand, personalized entertainment systems may be emerging as a key customer service for airlines despite the fact that investments of this nature have not materialized as market share drivers. Why? Such investments are not truly unique and, thus, can be matched by other airlines. In 2003, airlines globally spent $1.5 billion on in-flight entertainment and communications products and services, according to the World Airline Entertainment Association.

With regard to value chain analysis, even Southwest Airlines, which has used its value chain construction to derive significant comparative advantage, faces several challenges as its labor costs (in 1997, 45 percent of the total cost) are troubling for a low-cost airline. In addressing cost pressure, Southwest is focusing on growth by

---

[6] Mike Partridge and Lew Perren, "Assessing and Enhancing Strategic Capability: A Value-driven Approach," *Management Accounting*, 72, 1994: 28–30.

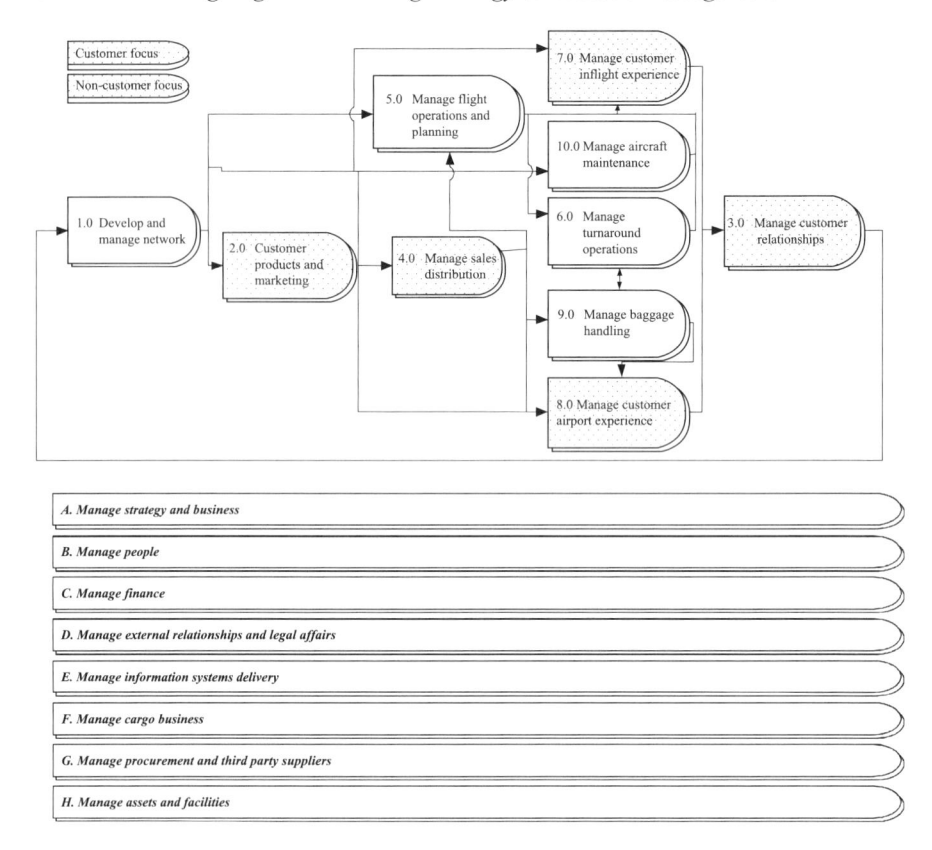

84   *Designing and Executing Strategy in Aviation Management*

Figure 5.2   Core Processes in a Passenger Airline Business Model

seeking longer routes and deploying cost-saving technologies. In 2003, Southwest's capacity grew just 4 percent, the lowest rate in 25 years. In 2004, it projected an 8 percent growth, followed by 11 percent in 2005. In terms of labor costs, Southwest is adding new workers at the bottom of the pay ladder. The lower-cost workers help offset the high costs of the airline's maturing workforce. Selling seats and getting them off the ground is expensive, but if Southwest can succeed in offering longer routes, the average cost per route will decrease. In 2004, Southwest expanded service to Philadelphia, and four of its six initial routes from this expansion airport are long distance:Phoenix, Las Vegas, Tampa and Orlando. In the future, many of the Philadelphia routes will be more than 700 miles. Regarding the acquisition and deployment of new, cost-saving technologies, it is interesting to point out Southwest dismisses their competitors' reliance on computers and software as wasteful spending. Southwest, like other low-cost airlines, does not sell tickets through the costly computerized reservations systems used by travel agents. Instead, it relies on its Internet site, a significantly lower-cost channel than travel agents. Southwest is also embracing new operational technologies in the air. For example, Southwest is acquiring new aircraft with "winglets," which make them more fuel efficient even though these planes come with a higher price tag than aircraft of the same type without the winglets.[7]

In analyzing value chain activities, it is important to look for value-creating activities at the different levels of the chain. Upstream activities, for example, are those activities involved in procurement, warehousing, and dissemination of inputs. In other words, they are the activities that are closer to the point of manufacturing or production. A firm might be able to cut costs, thereby adding value, by outsourcing previously in-house produced activities or looking for different suppliers. Likewise, cost-saving or value-adding activities might be found during the actual production or service cycle. For example, activities previously done separately can sometimes be accomplished in tandem, thus shortening production time, or, by cross training employees, there is less down time and the ability to speed up the production or service cycle. This approach was taken by People's Express in the 1981. People's Express was a pioneer "no frills" airline in the U.S., and by the beginning of 1986 it had become the fifth largest airline in the U.S. The company philosophy was that employees would be thoroughly cross-trained, thus shortening the service cycle time and reducing the cost of more specialized activities. This worked for People's Express to a certain extent but, as the airline found itself on the verge of becoming big, it saw that it had stretched too far. Its systems could not handle the rapidly increasing volume. In addition, People's Express also did not account for the fact that some of its employees did not want to be cross-trained or perform duties outside of their area of expertise. Consequently, People's Express went out of business in 1987 because it could not sustain growth and was purchased by Continental Airlines. The key is not what an airline might be next year but what it will be five years down the road.[8]

Similar to the upstream activities, the downstream activities of the value chain are those activities involved in getting the product to the customer. In other words, they are the activities closer to the point of sale. Some companies, like airline manufacturers, base production solely on "made to order" to reduce any unnecessary inventory costs. Value chain helps a company to look at its downstream activities and find ways where it can either reduce costs or improve value. Also included in downstream activities are those functions involved in selling and marketing the product. Assuring that marketing dollars are aimed at the correct target market is essential to operating efficiencies. Likewise, a company's approach to service can be an important value chain activity.

Table 5.1 shows an example of the use of value chain integration in several functional areas within the airline's operational and management structure via the use

---

[7] See <http://www.southwest.com/about_swa/airborne.html>.
[8] Michael Beer, "People Express Airlines: Rise and Decline," *Harvard Business School Case*, Harvard University Press, 2003, 1–23.

of information technology. The table shows the components of an integrated software architecture system and the business activity areas that this system supports.

Table 5.1 Value Chain through an Integrated Software Architecture

Important Components of the System

Airline Software and Technology Support
- Architecting the Information Technology (IT) Strategy: Successful strategy demands experience, objectivity and the ability to integrate hardware, software, process and organization. Starting with an understanding of its business objectives, an airline should evaluate the business landscape – market, industry, customer and competition.
- It should also analyze its use of technology, business application systems, organization, technical infrastructure, technology service providers and business workflow. It should identify what systems it needs to achieve its goals and what are its opportunities for improvement. Furthermore, it should evaluate the implications of change and develop or outsource IT and Information Systems strategy, which are tailored to the airline's specific business needs and goals.

Airline E-commerce and E-business
- B2b, b2c, and b2e E-commerce systems.
- Architecting and Implementing Airline E-business Strategies.
- Airline Booking Engines and Online Reservation Systems.
- E-CRM and Related Software.
- Computerized Reservation Systems (CRS) and Related Airlines Reservation Systems.

Reservation and Ticketing Systems
- Reservation System and Reservation CRS Based Services and Technology.
- Application Development and Maintenance for Transaction Systems.
- Solutions for GDS-Airline Interactions. EDIFACT Technologies, Web Services, XML-based technologies and related technologies.

In addition, enhancements, maintenance, and support can be provided for the following functions:

- Aircraft Engineering, Repair and Maintenance (MRO) Systems.
- Airline Payload Software and Systems.
- Airline Planning, Staff Scheduling and Tracking Systems.
- Airline Crew Management Software.
- Airline Costing and Accounting Systems.

Table 5.1 (continued)

- Airline CRM.
- Airline Cargo Software and Systems.
- Cargo Schedules, Booking and Tracking Software.
- Cargo Capacity Control Technology.
- Cargo Terminal Handling Software.
- Cargo Revenue Accounting.
- Airlines Messaging and EDI Software.
- ULD Control Technology.
- Cargo Claims Management Software.
- Cargo Communication Systems.
- Cargo Planning and Control Software.
- Aircraft Inventory Control Systems.
- Airline Information Processing and Distribution System.
- Airline Decision Support Systems (DSS).
- Alliance evaluation.
- Asset management.
- Network and schedule simulations.
- Fleet forecasting and valuation.
- Pricing, fuel price hedging.
- Airline and airport reports.
- Airline Market Intelligence Analysis Systems.
- Airline Corporate Information Systems.
- Airline Passenger Processing Software.
- Baggage Sorting, Tracking and Tracing Software.
- Check-in, Boarding and Security System.
- Departure Control Systems.
- Flight Firming Reservations Systems.
- Ticketing and Fare Quotes.

Through the software application depicted, an airline ties several key operational functions together. The system functions strategically to support its various operational activities and business functions from its e-commerce and its e-business activities to its reservations and ticketing systems in a way that all functional areas "talk to each other." In other words, they are tied to the same information technology platform and are effectively supported by compatible hardware and software. In addition, through the integrated software architecture system just described, enhancements, maintenance, and support can be provided for all operational and managerial functions within an airline, ranging from functional activities, such as costing and accounting, to engineering functions. The list of all the possible different areas of coverage are found in the latter part of the table.

What are the Problems with Value Chain?

To implement value chain fully, the traditional corporate barriers must be broken down, and this is not always easy or welcomed! In the traditional corporate structure, each area of the business is considered a separate entity, which does not necessarily work together or, in some cases, even talk to other areas of the business. Many companies have had this structure so engrained in their culture that the idea of trying to integrate functions seems almost impossible. Frankly, many companies do not have the culture for such an integrated approach. Nevertheless, if properly implemented, value chain can save a company on inventory and transaction costs and allow them to keep a closer eye on the suppliers and the customers. In a truly integrated company, many of the ideas for service and product development come from the customers.

Strategic Cost Analysis

A byproduct of the value chain is the strategic cost analysis. After the major elements of the value chain are identified, the next step is to break down the cost by assigning specific costs to specific activities. The purpose in doing this is to make cross company comparisons of the cost of the different activities. For this reason, the exact way the cost structure is broken down is dependent on how easy it is to get this cross company cost data.

Therefore, the way costs are analyzed is through what is known as activity based accounting. Rather than looking at traditional categories of expenses, such as total wages and benefits, materials, and R&D, with activity based accounting expenses are assigned based on the activity that is performed. For example, the wages and benefits of the individuals performing a given task would be assigned to this activity. By doing this it is easier to look at the activities involved in providing a service or producing a product and make comparisons with rival companies. Since so much of competition involves finding better ways to keep costs down, it is beneficial to see where costs exceed others in the industry. While one company might have an overall cost advantage over its rivals, strategic cost analysis helps to see if that cost advantage exists at all stages of production or service.

Cost comparisons across the industry look at things like how much inventory is maintained, how materials are purchased, as well as how quickly the company is getting the product to the market. Through strategic cost analysis, one company's management can learn from competitors.[9]

---

[9] Porter, 97–115.

Resource-Based View

Another way to look at a company is via the resource-based view. The resource-based view revolves around the idea of looking at a company in terms of the resources that are deployed. The idea of resource is broadly defined as "anything that can be defined as a strength or weakness" of a given organization. More narrowly defined, a resource is considered to be something that is fixed, something that cannot be altered easily in the short run, such as capital assets or a physical plant.

Let us look at this in terms of airlines. What is the most fixed factor an airline has? The fleet of airplanes, of course. If an airline can manage to keep its fleet or airplanes in the air longer than its competitors, then it is getting the most out of its resources. It is this resource-utilization that is a competitive advantage for the firm. Consider Southwest Airlines, once again. Most studies report that Southwest flies its planes on average 11.5 hours per day, compared to 8.6 hours for the industry. Already one can see an advantage. How do they do this and still have such a heavy schedule of short trips? Southwest avoids congested airports, helping them improve their turnaround times. In addition, their fleet is standardized, composed entirely of Boeing 737s. All in all, their resource-utilization provides them a competitive advantage in the industry.

But resources are not just "things," people can be resources, too. In fact, Southwest employs more people to help out at their airport turnaround, relying less on information technology, than their rivals. Despite the pressure for quick turnaround, Southwest's employees are dedicated to one flight at a time. The cost of labor may be higher, but in the long run the company boasts good employee relations. In a related issue, firm specific resources (like the labor force in this case) may vary from company to company within an industry. As previously discussed, Southwest operates on a point-to-point basis, while the larger airlines operate via hub-and-spoke. The hub might be a very valuable resource in the case of larger airlines.

The resource-based view suggests that the competitive history of the firm is important. For example, if a company is the first one to market a product, they may be able to hold on to this competitive advantage simply because they were the first ones to do so. But the resource-based view does not look at long-term sustainability. Yes, it does a good job of explaining how added value can be sustained, even in the event of imitations, but the long term viability of the strategic move is less clear. While the history provided by the resource-based view should not be discounted, in today's competitive marketplace it should only be thought of as a resource and not an end-all to strategic determinations.[10]

---

[10] Pankaj Ghemawat, *Strategy and the Business Landscape*, Reading, MA: Addison-Wesley: 1999, 116–119.

Strategy in the Twenty-first Century

Benchmarking: Learning from Others

To survive in today's fast paced world and prepare for the next generation, organizations are forced to rethink every aspect of their business, including structure, products, processes and markets. They must rethink the business to the point that will result in bringing their goods to the market more quickly, being more customer focused, innovative, flexible, and, in general, being able to handle change. Sometimes it is not necessary to "reinvent the wheel." In fact, some of the best practices an organization can adopt are not always new revelations. Successful organizations look outside their own industries to find best practices.

Benchmarking, as it is called, is the "practice of deriving important lessons from the experiences of other organizations."[11] Yet benchmarking is far more than just "surface copying." Benchmarking means thoroughly understanding the products, services and work processes involved in providing the service, from inception to delivery. It is these processes that are borrowed from others. For example, the Ritz Carlton has been the showcase of the hotel industry for its seamless service delivery. Numerous industries have looked to the Ritz to see what makes their service delivery so special.[12] Today, benchmarking is considered a universally acceptable business practice and has been applied successfully to nearly every aspect of an organization, from research and development to marketing and sales to production and distribution.

When an organization commits to benchmarking, it is committing to becoming world class. Benchmarking is living up to a standard of excellence. To benchmark, a company has to live up to certain performance measures, answering questions like, "How many?," "How quickly?," "How high?" and "How low?." But a benchmark is merely a fact of information. Knowing a benchmark does not automatically make a company world class. A business has to focus on implementing the improvements and, in order to do this it has to go out and figure out what makes other organization's work. As noted by Jarra and Zairi, when ICL Design and Distribution Ltd. wanted to improve its distribution system, it looked to Marks and Spencer.[13] Likewise, when Motorola wanted to speed up its delivery process of its cellular phones, it looked in the direction of Domino's Pizza and Federal Express.

The airlines have tended to look inside the industry to find best practices, such as American and its scheduling system, but the company that can think beyond the industry boundaries is the company that will set the standards for the next century.

---

[11] Yasar Jarrar and Mohamed Zairi, "Future Trends in Benchmarking for Competitive Advantage: A Global Survey," *Total Quality Management*, 12, 2001: 906–913.

[12] Bruce Schoenfeld, "Grounded In Service – Hotels Have Been Successful at Making Travelers Feel Good about Brand Loyalty," *Cigar Aficionado*, July/August 2002, 66.

[13] Jarrar and Zairi, 906–913.

Focusing on the Customer

Successful airlines are finding that focusing on the customer is essential to operational success. Frankly, travelers today are much more sophisticated than they were 10 or 15 years ago. Travelers want superior service or efficient, reliable transportation. There is little else in between. This is why concepts like benchmarking and value chain are so important. The best way for an airline, or any other company, to focus on customer needs is through a value-driven, flexible organization that revolves around process-based operations. An airline must first determine where it wants to position itself among customer needs and then build a capable organization around these needs. To do this, the organization must take a serious look at its internal environment and eliminate any duplication or costly, even valueless, activities. Sometimes organizations find that the best solution to keeping costs down and quality up is through outsourcing.

Outsourcing

Outsourcing involves the purchase of value-creating activities from an external source. In many industries, outsourcing is becoming more and more popular as a means of keeping costs lower by allowing the "experts" in a given area to provide a particular service or produce a particular product. For example, most airlines do not make their own meals for their flights. This activity is outsourced to a catering or restaurant business. Some automobile manufacturers outsource some of their key assembly line activities. Many other companies, including some airlines, have found that it is more efficient to outsource all of their computer programming activities. While the old philosophy was that an organization could keep costs down by performing all of their own services, we now know that companies can perform activities more cost effectively if the activity is something that they are truly good at doing. Virtually all companies today seek to find some value-creating activities from the outside. Few, if any, organizations have the resources or the capabilities to achieve superior competence in all of their primary and secondary value chain activities. By the same token, most companies cannot afford to develop all of the technologies needed to make them competitively successful.

Table 5.2 provides a list of areas where business process outsourcing is feasible. Some airlines, especially low-fare airlines, were created around the idea of outsourcing what they consider to be non-essential functions. This practice gave rise to the term "virtual airline." A virtual airline is one that has outsourced as many possible operational and business functions it can but still maintains effective strategic control of itself as a business unit. Outsourcing is done to reap as many economic benefits as possible, without eroding the core offering of the company. Some examples of areas where outsourcing can take place and produce positive results are finance and accounting, which are non-flight operations essential areas;

reservations processes, which would be very cumbersome, time consuming and costly for an airline to develop from scratch; and baggage handling, which can be done more cost effectively and equally efficient by a handling agent. Some airlines can also be virtual to the extent that they only lease aircraft and become established for short interim periods using the Air Operator Certificate (AOC) of another operator rather than their own.

Table 5.2 Examples in an Airline where Business Process Outsourcing is Applicable

Finance and Accounting
- Accounts.
- Accounts Receivable.
- Customer Billing.
- Audits.
- Revenue Accounting.

Back-office
- Refunds processing.
- Fares data entry.

Reservation Processes
- Ticketing queues management.
- Teletype rejects management.
- Pre-flighting.
- Flight Firming.

Baggage Services
- Handling in-bound phone queries relating to baggage that is missing, delayed, or lost.

Marketing and Customer Relationship Management
- Frequent flier program calls and service support:handling inbound ticketing and queries related to membership and working on code-share bookings for award redemption.
- Managing marketing, e-marketing and e-promotions programs for airlines.
- Research and business intelligence functions of Customer Segmentation.
- Data Mining and Analysis.
- Supporting an airline's marketing department in activities such as direct response functions, customer surveys, etc.

Table 5.2 (continued)

HR and Benefits Management
- Employee database administration and management.
- Employee expense reporting.
- Employee benefits administration.

Customer Service and Customer Interaction
- Airline Disruption Handling: handling customer inquiries for cases where there have been service disruptions.
- Customer interactions via multiple media – phone, snail mail, web and email.

Research and Data Gathering
- Market Research: data collection, analysis and presentation and business intelligence data gathering for decision support.

Strategy and the Internet

There is no doubt the Internet is changing the way business is conducted throughout the world. The Internet allows businesses to compete more easily on a global playing field. E-commerce has changed the complexion of competition, adding new driving forces and new issues with which to contend. From the standpoint of the airlines, e-commerce has given more power to the customer. Customers can now book their trips, search for the best deals, compare prices, and, with a click of the finger, avoid all of the obstacles previously encountered in booking a trip. The Internet has provided quick response to the customer, thus empowering the customer even more.

Today, the availability of sophisticated computer systems and comprehensive databases covering a wide range of commercial activities and services together with satellite-based telecommunications facilities has had a significant impact on the aviation industry. The current worldwide network of efficient telecommunications facilities and services, especially the Internet, is providing the business community with substantial improvements in doing business on a worldwide level.

The world's leading airlines have achieved success as a result of a robust e-commerce strategy. Some of these successes include the following:

- Increased revenues at much lower costs per booking.
- Offering a diverse set of inventory in addition to air products, contributing to both higher customer satisfaction and increased revenues.
- Putting the control of the booking process in the traveler's hands and, hence, making the entire process simpler and more rewarding for both the traveler and the airline.
- Enhanced ability to issue e-tickets.

- Integration with their related databases, such as frequent flier program data and customer profiles data, to provide higher value to the traveler while at the same enabling the airline to have a more powerful Management Information System (MIS).
- Constant learning about the customer and his or her behavior and the ability to use this information almost instantaneously to provide higher value to the traveler.
- Extraordinary flexibility, be it in the booking process flows, presentation of information, or the ability to search on a range of travel-related parameters.
- Ease of scaling customer access to higher levels of service with corresponding increase in costs.

Knowledge Management

One of the important issues in business today is how information is managed and retained in an organization. The idea of knowledge management in an organization is simply making sure that managers have all the tools and knowledge needed to make decisions. However, some researchers think of this as a conundrum:if a firm is conscious to make sure that it has a specific process and that process is well documented, then it is certain that competitors also know the process! What should a company do? The process should be made easy enough for employees to understand yet tough enough that it is difficult for customers (or competitors) to copy.

Knowledge management is considered to be a critical issue for all businesses. In fact, it is really a critical success factor:knowledge must be made available to the right people at the right time, and, at the same time, it must be protected so that this knowledge does not leave the organization. What businesses are really competing for in the global world is knowledge and information. Companies have a vast capacity for knowledge, and what is important is that they learn to reuse this knowledge. Often a new strategy is nothing more than a re-packaging of something that some company has done before. The key is knowing what has been done before. Intellectual assets are very valuable to an organization. Processes or ideas dreamed up in the human mind today can be priceless tomorrow. Therefore, it is critical for a company to learn to manage intellectual assets, because this information can be the basis for strategic acts in the future.[14,15]

---

[14] T.H. Davenport, R.I. Thomas and K.C. Desouza, "Reusing Intellectual Assets," *Industrial Management*, 45, 2003: 12–19.

[15] Mutiran Alazmi and Mohamed Zairi, "Knowledge Management Critical Success Factors," *Total Quality Management and Business Excellence*, 14, 2003: 199–204.

Case Illustration: JetBlue: Value Added[16]

JetBlue can be cited as an example of value-based service. There is much that is distinctive about JetBlue, from the enthusiasm of its employees to its relentless customer focus to its chic, slightly countercultural image. These are precisely the sorts of things smaller organizations can achieve and which become far more difficult the larger the organization grows. Can JetBlue maintain these qualities as it grows into the bureaucracy that is required to manage a vastly more complex operation? Thus far, JetBlue has avoided the use of policy manuals. Whether valued service can be sustained is a valid and very difficult question that applies to many unique and innovative companies in several industries. Companies that can sustain their value-added outlook are referred to as "postmodern corporations." If they manage this transition, then they can grow, but they still remain, in important ways, the antithesis of bigness as they depend on flexibility, speed and a sense of intimacy with employees and customers alike. Is a business model as such scaleable? At JetBlue they call it a "value-based model" as employees (called crewmembers) are expected to rely on the company's core values in order carry out their jobs rather than following a set of standard operating procedures found in a policy manual.

For now, at the time of the publication of this book, JetBlue enjoys one of the lowest costs in the industry, just over six cents per passenger seat mile (see Table 1.1 in Chapter 1). But any startup with new planes will have lower costs than established airlines with older aircraft. Presently, JetBlue aircraft are not in need of heavy maintenance. Furthermore, none of the crewmembers have worked at the airline more than a few years, so their pay scale is fairly low. This will also change with time. Legacy airlines utilize more money to run their business.

David Neeleman, the founder and CEO of JetBlue, is not a big company kind of person. He attracted and adapted mature talent to a new company: Dave Barger, president and COO, ran Continental's Newark, New Jersey, hub; John Owen, CFO, was treasurer at Southwest; the head of Human Resources, Ann Rhoades, also came from Southwest; and Al Spain, Senior Vice President of Operations, had experience with People's Express, Continental and Continental Micronesia. Neelman's philosophy is that if exceptional talent is recruited and brought into a new company, they will add significant value as they are able to act independently and efficiently, producing desired outcomes.

Despite wanting to maintain a small company culture, JetBlue's management realizes they are competing both with big as well as small companies. For example, the company management must understand the processes and tools needed to drive consistency through the operation as the operation grows, the often mundane attention

---

[16] Al Spain, "The Business Strategy of JetBlue," lecture given at Aviation MBA Program at Concordia University.

to details and numbers that seem to elude so many entrepreneurs. Every morning members of the operations team at JetBlue review the previous day's flights. They review data, like takeoff and landing times and when the last bag hit the belt, having as their goal no longer than 20 minutes after the plane reaches the gate. As Al Spain noted, "The 45 minutes you wait for your bags is your last impression."

JetBlue's operations team also reviews a list of "focus flights," the ten worst delays. Since, on average, the flying public has low expectations of low-fare carriers, vis-à-vis service, JetBlue wants to differentiate itself through both reliability and service. Two of the most important performance numbers are the completion rate, the proportion of flights that are not canceled (in which JetBlue is an industry leader) and on-time arrivals, where JetBlue is, again, an industry leader.

One of the tools designed to help the company as it grows from an operations reliability standpoint is an "operational recovery system." During any operational disruption caused by weather, security concerns, or an accident, the system enables planners to select various goals before rerouting planes. The goal is to prevent canceled flights or flight delays beyond three hours. The software produces a solution and calculates the operational cost of the solution, factoring in each plane's maintenance and fuel needs and the flight crew's experience and availability within Federal Aviation Administration rules. With a fleet of less than 70 aircraft, the program is a nice addition to the operational system. However, if the fleet increased to 140 planes, it would be indispensable.

As they manage growth, airlines must also standardize many other functions they engage in to avoid re-inventing the wheel every time. For example, JetBlue has developed a checklist of what has to happen whenever it enters a new market. Everyone involved has access to the list on the company intranet. Each department sees what has been done, what remains to be done, milestones and problems. Currently, the checklist makes inauguration or the launching of new services, which occur months apart, more predictable. Before too long, it could make simultaneous launches quite normal and therefore manageable. The emphasis on efficiency improvement, should be critical in the years ahead as JetBlue tries to offset rising costs for aging planes and more-senior employees. Low costs remain an inextricable part of their business model. As an example of a cost cutting measure, JetBlue's reservation agents work from home rather than at a corporate call center. This practice is designed to save costs and keep people at home, where they feel more comfortable and can stay close to their families. On the other hand, the company may encounter problems with service reliability by using this practice, even though safeguards have been put into place to avoid such disruptions. Interestingly, in their effort to address profit margins, JetBlue departed from a conventional low-cost airline practice of operating a single type aircraft fleet, adding a new 100-seat regional jet fleet (EMB 190) to tap into relatively uncontested and, as such, theoretically more profitable markets in city pairs that cannot support service by larger jets, like the current one (A320) which constitutes JetBlue's fleet.

Can JetBlue feasibly achieve all this, grow, standardize, and automate, while still preserving its personal touch? According to Al Spain, the real secret weapon of JetBlue is the employees, all of whom are called "crewmembers" to emphasize the companywide sense of teamwork. According to JetBlue's philosophy, if you treat your crewmembers well, they will treat the customer well. "There is no 'they' here," says Spain, "it is 'we' and 'us'. We succeed together or we fail together."

JetBlue's staff is not unionized. The philosophy that the company espouses is that if management and crewmembers trust one another and if people feel they are compensated fairly, there should be no need for a third party, like a union, to intervene in the relationship. Also, crewmembers will be more likely to stick by the company in tough times if they have incentives, like profit sharing.

A further challenge for a fast growing company is the harmonizing of new and old employees, the individuals who are just joining the company and have not been exposed to the culture and the ones who have been with the airline for years. Both groups are continuously stimulated by making them realize how much their contributions matter. The pilot group is an interesting example of how JetBlue achieves this. At most airlines, pilots are viewed as one-dimensional technical experts. JetBlue, however, encourages its pilots to participate in the business itself. Some design airport diagrams to help orient colleagues; others help in financial analysis for the company; and still others make inventories of his or her fellow pilots' skills in hopes of identifying other abilities that might be useful to the airline. This practice resembles the cross-functionality model followed by People's Express. Will JetBlue be successful where People's Express failed? Only time will tell.

One of the main reasons as to why senior leadership at JetBlue has been effective at building a dedicated staff is that their visibility to the crewmembers is frequent. Senior management cultivate and champion the culture on the front lines: in addition to flying JetBlue most weeks, they appear at orientation meetings for new hires and conduct monthly "pocket sessions," informal Q&As with crewmembers. In essence, they practice what they preach, adding significant credibility to their message.

Chapter 6

Setting Corporate Direction

> Think and act big and grow smaller, or think and act small and grow bigger.
> Herb Kelleher[1]

Previous chapters have established the fact that organizations exist for a purpose (stated in their mission and vision) and are built on a set of core values. In these previous chapters, the idea of how core competencies are used to establish a corporate strategy and establish a sustainable competitive position was discussed. The essence of corporate strategy is determining the overall direction that will enable the purpose of the organization to be fulfilled. Corporate strategy achieves strategic objectives through the products or services the business offers. One aspect of corporate strategy that should reflect the purpose and core values of an organization is whether the organization should remain a single business or operate in multiple businesses.

What Will it be – Single or Multi-Business?

A single business strategy is exactly what it sounds like, focusing on only one business. A single business strategy can be a very attractive strategy for many organizations. When a company decides to embark on a single business strategy it is making the decision to try to be the best it can be, and all the resources of the company are directed toward perfecting that business. Efficiency of operations becomes a given, but efforts to re-engineer operations continually in order to improve operational efficiencies are generally the order of business. A single business strategy means that the company continually looks for next-generation applications.

A company can be a single business organization and still have multiple products, multiple markets and multiple outlets, as long as all remain in the same industry. For example, a cargo company may operate in the United States as well as in several other countries in several continents, each country having its own unique market characteristics. A real world example is Polar Air Cargo, a global carrier specializing in time-definite, airport-to-airport scheduled air freight service, which began round-the-world scheduled cargo service in 2003. Polar provides a critical link in the international logistics chain by connecting major cargo markets in Asia, Europe, Australia, and the Americas with frequent Boeing 747 freighter services.

---

[1] See <http://www.skygod.com/quotes/airline.html>.

With facilities at major United States' gateways, including New York, Los Angeles, Chicago and Miami, and extensive networks in Asia, including Tokyo-Narita and Hong Kong, Polar provides international freight forwarders and agents with reliable connections on scheduled routes. Polar concentrates only on scheduled air transport of cargo, but in many markets worldwide.[2]

Some believe that selecting a single business strategy is too risky. There are risks, of course. A single business strategy is like putting all eggs in one basket; if the basket drops, everything is lost. Likewise, if the industry collapses there is nothing else on which to fall back. Therefore, sometimes the best choice is a multi-business strategy.

A multi-business organization operates in more than one industry. For example, General Motors could be considered a multi-business organization, because it has business units in the automobile industry as well as the aerospace industry. A good example of a multi-business organization in the aviation industry is Lufthansa. They operate in several business sectors of the airline industry, using a strategy of "being the best you can be" in each specific sector. Lufthansa operates scheduled airline service in a global network, is a major partner in STAR alliance and its subsidiary companies, (e.g. Lufthansa Technik and Lufthansa Consulting) and partakes in the business of aircraft maintenance and overhaul and aviation business consulting. Lufthansa also operates regional airline services and exclusive executive type airline services.[3] Many people consider the benefit of a multi-business organization the fact that the company has spread its risk across different industries. As a result, there may be a downturn in one industry but the other industry may remain viable.

The decision as to whether a company remains a single business or branches into multi-businesses should be based on the mission and vision of the company. Remember, the mission helps to set the direction of future business decisions, and this is one of them. Whether the company remains a single business or a multi-business enterprise is important, because it further affects its overall strategic direction, what corporate strategies to follow and how to implement these strategies.

Corporate Strategies

The corporate strategy establishes the overall direction that a company wants to follow. This differs from the generic strategies discussed in Chapter 2. Competitive strategies help a company determine the means of arriving at a corporate strategy. For example, if the overall corporate strategy is one of international expansion, the resulting competitive strategies are built on core competencies to help enter into these

---

[2] See <http: //www.polarairways.com>.
[3] See <http: //konzern.lufthansa.com/EN/html>.

new markets by providing a product or service better than the competition in that market. A corporate strategy is dependent on competitive strategies and can take on several different forms. First, let us examine growth strategies.

Growth Strategies

Many people believe that unless a company is undergoing some type of growth strategy it will never be successful. While this idea is open to debate, nonetheless, a growth strategy is certainly an attractive strategy for any organization. The purpose of a growth strategy is to increase the operations of the organization by some means, which could mean entering into new markets or simply attracting more customers. It could be increasing sales or increasing the product or service offerings. Whatever method is taken, it should increase some aspect of the business. Growth strategies can be classified as intensive, integrative, concentration, or diversification.

Intensive Growth Strategies

An intensive growth strategy works best when the industry the business operates in is a growing industry. This means there are new customers, markets, or even products yet to be tapped. A company might embark on one or more intensive strategies. These include, market penetration, market development, and product development.

Market penetration Market penetration means that a business remains in the same market but wants to attract more customers. In other words, the goal is, in essence, to steal customers from a competitor or convince customers in the market who have never tried a product or service to do so. How is market penetration achieved? It may be done by choosing to invest more in advertising or more in sales force. The company might offer special promotions or discounts. Airlines are notorious for implementing market penetration strategies when they offer special prices on travel to particular destinations. Canada's JetsGo engaged in such a penetration strategy in April 2004 by promoting a fare sale of $1 CAD, in the Montreal to Toronto city pair. The overall ticket cost, including taxes and fees, was about $80 CAD, which was still a bargain when compared to the prices of incumbents, such as Air Canada, in the same city pair. JetsGo embarked on an aggressive expansion strategy in the spring of 2004 with the purchase of used F100 aircraft from American Airlines, more than doubling its fleet size. JetsGo decided to challenge Air Canada and WestJet in some very lucrative routes, for example, the Montreal to Toronto shuttle where, in its estimation, the markets had not reached their full potential and were under serviced and overpriced. Although its unorthodox strategy drew criticism from both Air Canada and WestJet executives, JetsGo's move was in line with a market penetration strategy. JetsGo increased its presence in the Montreal to Toronto city pair from a couple of flights a day to hourly flights. No matter what method is used to implement a market penetration strategy, the objective is to increase market share in an existing market.

Market development When a company's management team selects a market development strategy, they have decided to take their product(s) or service(s) into a new market. The new market can be a geographically different market, such as overseas or international expansion, or it can be a new customer market. For example, maybe the product or service has only been directed at the adult market up to this point. Using a market development strategy, the company might choose to offer products and services in the children or adolescent's market.

Spirit Airlines provides an example of international market development. With $125 million USD in new capital and new owners, it announced in February 2004 its intention to expand into several international routes and challenge major airline competitors. Spirit operates out of Detroit Metropolitan Airport and is based in Miramar, Florida. In February 2004, Spirit secured government approval to fly to 11 countries. Until then, it had one international flight, to Cancun, Mexico, out of Ft. Lauderdale, Florida. The new plan was to fly to more international destinations from Fort Lauderdale and begin international service out of Detroit. The 11 countries approved were Aruba, the Bahamas, Canada, Costa Rica, the Dominican Republic, El Salvador, Guatemala, Honduras, Jamaica, Nicaragua and Panama.

As a result of this decision, Spirit was projecting its revenue to grow from $450 million USD in 2003 to about $1 billion by 2008. Spirit's move came as other low-fare carriers were expanding to international destinations, such as ATA Airlines, which even announced low-fare airline service to Europe.[4]

Product development When a company's management team selects a product development strategy, they are making the decision to launch a new product in their existing market. The new product might be a complementary product, or it might be a less expensive or more upscale version of the existing product. The product development option creates a new product for use by the company's current customers. Delta's "Song," the "low-cost," "no-frills" derivative of Delta AirLines, is such an example. Delta, although unsuccessful with its Delta Express idea, decided to launch another version of a low-cost carrier, Song, using a very similar business model to that of Delta Express. Song revised its original strategy of concentrating on the East Coast's North-South corridor and started competing primarily with JetBlue on East-West routes. JetBlue proved that there was strategic potential in East-West markets, and Song tried to make sure that, where Delta was not well suited to compete, it would try to do so on its behalf. Song was to serve, primarily, the very busy Northeast-to-

---

[4] James Wallace, "Aerospace Digest: Low Fares to Europe May be Coming," *Seattle Post*, 18 February 2004, see <http://seattlepi.nwsource.com/business/161034_air18.html?searchpagefrom=1&searchdiff=482>.

Florida corridor, with high frequency direct flights. Since these direct flights would bypass Song's traditional "eastern" hubs of Atlanta and Cincinnati, it would, thus, offer an alternative airline to true breed low-cost operations, such as JetBlue. As a "product" of Delta, Song is able to take advantage of privileges offered through Delta, such as technical operations, training, frequent flier programs, and reservations systems. At the same time, Song wants to avoid other typical "legacy costs" associated with Delta. Typical legacy costs, for example, are those associated with the productivity of an airline's workforce, which can directly contribute to the company's operating margins. Other mainline operations have tried to offer the same type of services to its consumers, such as Ted (United's low-cost offering) and Tango (Air Canada's short-lived, low-cost subsidiary). The possible reasons as to why these mainline carriers have used low-cost subsidiaries to compete are twofold. First, they may be trying to beat low-cost carriers at their own game of simplified price service offerings and lower prices and, at the same time, offer some "legacy" services to their customers (e.g. global network, frequent flier programs stretching to their alliance partners, etc.), or, secondly, they may be trying to create a test base, which they can use to transform their mainline operation into a low-cost one.

Integrative Growth Strategies

Integrative growth strategies are most effective when the marketplace is beginning to reach saturation. Integrative strategies would not make sense in a new and growing market because there is plenty of market share available to all players. However, when the marketplace begins to get crowded and there is less and less opportunity to find pockets of untapped markets, integrative strategies work best. There are two kinds of integrative strategies, horizontal integration and vertical integration.

Horizontal integration Using a strategy of horizontal integration, growth occurs by purchasing competitor firms. The object is to gain greater market share by expanding the reach of the company while at the same time eliminating competitors. This keeps the company in the same industry but allows it the chance to strengthen its position in that industry. However, there is one drawback to horizontal integration: the threat of antitrust laws. If joining together with another company can be perceived to cause a monopoly-type situation in a given industry, the Federal Trade Commission and the Department of Justice will impose antitrust regulations and block the merger. Basically, if the merger causes competition to diminish greatly and, therefore, causes a lack of choice for the consumer, it is likely the integration of the companies will be prohibited. Assuming the merger of the companies is considered to be legal under antitrust laws, horizontal integration makes sense as long as the combination allows the originating company to attain or maintain a sustainable competitive advantage. It is important that all aspects of the merger be examined, from operating systems to corporate culture.

There have been several examples of horizontal integration, or horizontal integration attempts, in the airline industry both in the United States and in Europe: United's attempted buyout of US Airways; Trans World Airlines' (TWA) purchase of Ozark; US Airways' purchase of independents (Allegheny, Mohawk, PSA and Piedmont); Air France's acquisition of KLM; and American Airlines buyout of TWA.

The American Airlines-TWA story is fascinating given the timeframe in which it took place, that is, a few months before September 11, 2001. On January 8, 2001, the *Wall Street Journal* reported that American Airlines offered to purchase Trans World Airlines, effectively taking the troubled carrier out of Chapter 11 bankruptcy. The announcement began a public debate about whether or not the merger was a sound business decision that would ultimately benefit the shareholders of American Airlines. From the beginning, financial markets did not believe that the acquisition of TWA assets was a value enhancing decision or that there were significant strategic benefits from the merger. Equity value declined more than 30 percent because of the takeover.[5] The events of 9-11 further changed the economic landscape of the industry and made it even more difficult for American to survive. Following September 11, 2001, American Airlines substantially underperformed in comparison with other major solvent airlines in the United States.[6] The events of the merger, and then 9-11, eventually led to the resignation of CEO Don Carty and American embarked on a cost-cutting plan.

Vertical integration Vertical integration is often known as backward and forward integration. The goal of vertical integration is for the company to gain control of inputs or suppliers (backward), outputs or distributors (forward), or both. Vertical integration is considered a growth strategy because it expands the organization's operations while remaining in the same industry.

When a company embarks on a strategy of backward integration, often called upstream integration, it becomes its own supplier to at least one of its component parts. For instance, Pratt and Whitney became its own supplier of airline components when it realized that it could beat both the quality and price of component parts.

When a company chooses a forward integration (downstream integration) strategy, it becomes its own distributor. This may be through retail or outlet stores or through franchising. Or, for example, a company may own a trucking firm that ships its products to the end user.

What are the benefits of vertical integration?

- Reduced purchasing and selling costs.

---

[5] For a thorough analysis of the financial results see Triant Flouris and Steve Swidler, "American Airlines' Takeover of TWA: An Ex-Post Analysis of Financial Market Information," *Journal of Air Transport Management*, 10, 2004: 173–180.
[6] Ibid.

- Improved coordination among functions and capabilities.
- The ability to protect proprietary products.

What are the concerns of vertical integration?
- Reduced flexibility.
- Difficulty in integrating the operations.
- High cost of acquiring a firm or starting from scratch.

Concentration Growth Strategies

A concentration strategy is a type of growth strategy where the company concentrates on its primary line of business, looking for a means of increasing its level of operations in this business. The benefit of this strategy is that the company remains in its core industry. This strategy is often used by a company choosing to remain in a single business. As long as the growth strategies are being met, it is an attractive strategy to undertake.

How does one accomplish growth through a concentration strategy? A concentration strategy means the company concentrates all of its efforts on improving operations. The company looks for ways to increase business through improved products and services. It could even include taking a new product to a new market, as long as the company does not stray outside of its core industry.

Similar to the argument for a single business operation, a concentration strategy allows the company to become very good at what it does. It allows the company to know the market and competition inside and out and fine tune all of its operations. No one in the organization is spread too thin in their knowledge of the business and everyone's goal is to exploit the company's core competencies. However, like the argument for remaining a single business, a concentration strategy can open the company up to the vulnerabilities of the market, leaving it with nothing to fall back on if there is a downturn or total evaporation of an industry. The latter could be the result of technological advances that eliminate the need for products in a given industry. The key to implementing a concentration strategy is to be continually aware of the external environment and its opportunities and constraints.

Diversification Growth Strategies

When a company selects a diversification strategy, the company is choosing to enter into different products and different markets. In other words, the company grows by moving its business operations into other industries. One of the motives for diversification that has been suggested is the transaction cost.[7] In other words, the company decides to use its excess resources and, therefore, enhances efficiency.

---

[7] R. Amit and J. Livnat, "A Concept of Conglomerate Diversification," *Journal of Management*, 14, no. 4, 1988: 593–604.

There are two basic types of diversification, related and unrelated. Related diversification means the company is expanding into another industry but one that is similar to that of its core product. Unrelated diversification simply means that the company is expanding into a totally unrelated or different industry.

Related diversification Related diversification means that the company diversifies into an area that will create some kind of strategic fit, achieving synergy. Related diversification generally occurs when industry profitability is increasing. At this point, management decides to embark on related or offensive diversification. This type of diversification refers to the ability to exploit the resources or capabilities in the firm's current industry. There should be some overlap either in resources used (materials or human), markets served, similar products, similar operational needs, or distinctive competency. This overlap means that a company does not completely duplicate all aspects of the business, creating some kind of cost savings. It is the old idea of $1 + 1 = 3$. The company gains more from the combination than from operating both businesses separately.[8]

Related diversification can be divided into two categories, horizontal and concentric.

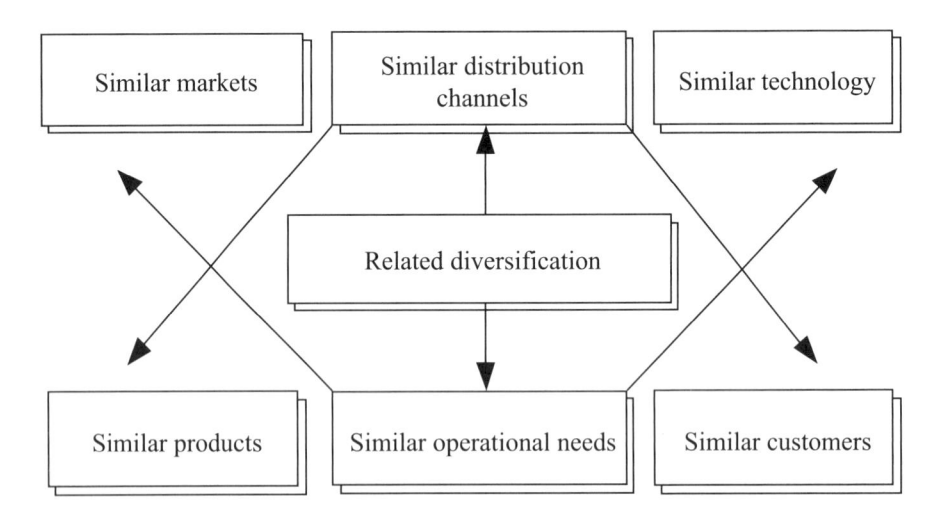

Figure 6.1   Related Diversification

---

[8] Choelson Park, "The Effects of Prior Performance on the Choice Between Related and Unrelated Acquisitions: Implications for the Performance Consequences of Diversification Strategy," *Journal of Management Studies*, 39, no. 7, 2002: 1003–1019.

- *Horizontal Diversification* When a company diversifies horizontally, the company is expanding operations by producing a new, completely different product or providing a new, completely different service in a different industry for the same class of people. In other words, the company wants its existing customers to purchase this product or service. For example, an upscale airline might decide to enter into the upscale hotel business. The customer profile for both services would be affluent individuals, but the industries for these services are completely different but still related.
- *Concentric Diversification* Concentric diversification relates to expanding operations by producing the same product or service for a completely different class of people. In other words, the company is looking to expand its business reach to new people, people who were not already using its products. For example, an upscale, full-service airline dedicated to the business traveler and luxury traveler might decide to begin operations of a no-frills, bargain airline dedicated to family travelers.

United Parcel Service: Related Diversification

United Parcel Service (UPS) provides an example of related diversification. UPS is a leading contender in the package delivery arena and a global leader in supply chain services. There are several component companies to UPS: UPS Air Cargo provides direct, airport-to-airport deliveries to over 150 airports (as of 2004) around the world and is one of the world's largest all cargo airlines; UPS Capital provides businesses with working capital solutions that integrate supply-chain management with supply-chain financing; UPS Consulting specializes in supply chain management consulting; UPS Mail Innovations handles mail communications by providing upstream mail processes (sorting, weighing and applying postage) and distributing them to the U.S. Postal Service for final delivery; Mail Boxes Etc. (owned by UPS since 2001 and also branded as "The UPS Store") is a franchisor of retail shipping, postal and business supplies; UPS Supply Chain Solutions encompasses UPS Logistics, UPS Capital, UPS Freight Services, and UPS Mail Innovations; and UPS Professional Services, Inc. (PSI) focuses on strategic business solutions for businesses. All of these businesses, while different, are related, providing synergy within the company.[9]

Unrelated diversification As previously mentioned, unrelated diversification occurs when a company expands operations into an industry that has no relation to the core industry, for example an airline diversifying into a chain of hospitals. In this

---

[9] See <http://www.ups.com/content/corp/index.jsx>.

instance, there is no strategic fit, no synergy. Basically, there is no compatibility with the company's core competencies, and there is no cost advantage to entering into this industry. This seems to go against everything that has been discussed so far in strategy.

Why would a company diversify into unrelated businesses? When an industry has low profitability and is structurally unattractive, the management team may pursue unrelated diversification, which is often referred to as defensive diversification. Defensive diversification avoids the adverse environmental factors in the company's traditional industries. The motive, here, is clear: management wants to improve the overall prospects for the firm. Since the firm is in a structurally unattractive or low-profit industry, it would not make sense to diversify into related industries because it is likely that they will also be unprofitable and structurally unattractive. Consequently, firms in low-profit industries tend to diversify into unrelated businesses. It is a way of spreading the risk outside the existing industry.

What are the pitfalls? Without synergy, there seems to be no chance of profitability above what the businesses could produce if they were independent of each other. Further, without expertise in the new area, management will either have to be hired, adding to costs, or the current management might have to "learn the business quickly."

Everything being equal, unrelated diversification is a greater risk than related diversification. Why? Because it represents a greater deviation from the current industry in terms of products, markets, technologies and even ways of conducting business than one would find with related diversification. As a result, it is often more difficult to predict future performance with unrelated diversification than with related.[10]

Joint ventures and strategic alliances Joint ventures and strategic alliances are common strategies undertaken in businesses in recent years. A joint venture means that two or more companies pool their resources to undertake a mutually beneficial project. This is common in the semiconductor industry. A joint venture allows companies to share costs, resources, and technology. It also reduces risk by spreading cost of capital among a number of firms. One example of this is the joint venture between Hitachi and Mitsubishi. The 50/50 joint venture, with expected revenues of $5.3 billion, encompasses all aspects of chip development from R&D to manufacturing. The benefit is expected to come in cost reductions and development speed. The merger created the industry's third-largest chip producer after Intel and Toshiba."[11]

Another example is the joint venture among five airlines creating the Orbitz online travel consortium. By pooling financial capital and industry expertise from American, Continental, Delta, Northwest and United, Orbitz allows users to scan more than two

---

[10] Ibid.
[11] See <http://www.semi.org/web/wmagazine.nsf/>.

billion fare possibilities from over 450 airlines in seconds. Travelers should get lower air fares in a more convenient medium and airlines counter the threat of Expedia, Travelocity, Priceline.com and other sites like them at little risk.

A strategic alliance is different from a joint venture in that companies typically operate under an agreement to share resources rather than commit to an entirely new project. While there are several strategic alliance teams in the world, Northwest's alliance with KLM was interesting to the degree in which mutual cooperation occurred. All seats were jointly marketed and, while for legal reasons they were two independent companies, to a reservations agent the airline looked uniform. The airline industry calls this type of cooperation codesharing.[12] For example, Northwest Airlines flight #459 would also be called KLM flight #8578; yet, the plane, crew and gate would be Northwest's. However, in ticketing and revenues, the traveler flying on flight #459 would have purchased a KLM ticket. All planes have each other's logo, and all websites have each other's name, logo and products. The cooperation was seamless, and the benefit was that more destinations and frequency of service were available through the allied marketing of route structure.

The alliance between KLM and Northwest, which was known as Wings, was the result of the first open-skies bilateral agreement, signed in December 1992 between the United States and the Netherlands. The alliance subsequently grew to include Continental Airlines and then was terminated with the acquisition of KLM by Air France. This alliance has historical significance vis-à-vis international air transportation as regulators in both countries afforded both carriers antitrust immunity, giving the airlines the ability to establish joint fares and schedules, which was the first time such permission had been given.

Strategic alliances are of significance and value in the airline industry, as airlines have been seeking various ways of increasing their strength in the marketplace through mergers, the purchase of equity in other carriers, and a variety of joint marketing agreements and cooperative activities. In the last two decades, and as a result of more deregulated business environments, codesharing has become the basis of many strategic alliances for the world's mainline carriers and avoided by low-cost carriers. Codesharing enables airlines to expand their networks. Airlines throughout the world have concluded a wide range of alliance and partnership agreements, covering many areas of airline activities, services and operations. Initially, the main emphasis was directed toward codesharing, block seat and space agreements, coordination of flight schedules and frequent flier programs.

Now, alliances have evolved to include the possibility of joint aircraft purchasing. They also cover many other fields of commercial, technical and operational activities,

---

[12] The basic concept of codesharing is that passengers fly a segment of their journey on an airline other than the airline that sold them the ticket for that particular flight. Codesharing is derived from the concept found in the IATA Multilateral Interline Traffic Agreements called interlining.

such as marketing, advertising, ground handling, catering and aircraft maintenance. In some instances, airline cooperation extends to sharing terminal facilities for both passenger and cargo, sharing airport slots, joint fleet planning and joint purchasing and sharing of spare parts for aircraft and engines. Airline alliance and partnership agreements differ considerably, both in nature and purpose. Most strategic alliances involve an agreement between two or more air carriers designed to create a global route network with integrated marketing, distribution, and other services.

As of 2005, there are three main strategic alliances. Star Alliance, founded by Lufthansa German Airlines (Germany) and United Airlines (USA) in May 1997, includes Air Canada (Canada), Air New Zealand (New Zealand), All Nippon Airways (ANA-Japan), Asiana Airlines (South Korea), Austrian Airlines (Austria), bmi-British Midland (Great Britain), LOT and Polish Airlines (Poland), TAP Air Portugal (Portugal), Scandinavian Airlines System (SAS, Sweden, Norway, Denmark), Singapore Airlines (Singapore), South African Airways (South Africa), Spanair (Spain), Swiss International Airlines (Switzerland), Thai Airways (Thailand), US Airways (USA) and VARIG (Brazil). A major long-term objective of STAR is to share the same terminal buildings in key airports around the world as a means of cost streamlining. On January 1, 2002, the alliance took corporate form as Star Alliance GmbH under German law. Its headquarters are in Frankfurt, Germany, and it is considered the most integrated of the alliances. Alliance members have even considered joint purchasing and operation of aircraft. Several aircraft owned by STAR Alliance members are painted in Star Alliance livery, which includes the trademarks of several members of the alliance and the Star Alliance Trademark (STAR) on the tail.

Oneworld was founded by American Airlines (USA) and British Airways (Great Britain) in September 1998, and includes Aer Lingus (Ireland), Cathay Pacific (Hong Kong), Finnair (Finland), Iberia (Spain), LanChile (Chile), Qantas (Australia) and several codesharing partners. An important activity for Oneworld is the joint procurement of parts and services in areas such as engineering and maintenance, based on standard specifications. This joint procurement is done in an effort to streamline these operations and, thus, produce savings for its member airlines.

SkyTeam was founded by Air France (France) and Delta AirLines (USA) in June 1999, and includes Aeromexico (Mexico), Alitalia (Italy), Czech Airlines (CSA, Czech Republic), Korean Air (South Korea), Northwest Airlines (USA), Continental (USA) and other codesharing partners.[13] In September 2001, SkyTeam Cargo was launched comprising AeroMexico Cargo, Air France Cargo, Alitalia Cargo, CSA Cargo, Delta Air Logistics and Korean Air Cargo.

---

[13] Air France merged with KLM on May 5, 2004. On September 5, 2004, Air France transferred its assets, liabilities, and business operations to a wholly-owned French subsidiary and was renamed Air France-KLM. The new combined airline is a member of SkyTeam.

No Growth Strategies

Retrenchment and Turnaround

Turning around a failed business may sound exciting and, from a business sense, maybe at bit romantic. Unfortunately, it is not as exciting as it sounds and it does not always work. There is no single formula or even a single timescale to assure successful turnaround, but there are some common themes. Let us begin with the whole premise behind retrenchment and turnaround.[14]

Retrenchment means to cut back. A retrenchment strategy is often used when a worthwhile business goes into crisis. First, management has to determine the root of the problem. Is it the result of a poor economy that has caused a downturn in sales? Is it because of a poorly concocted strategy? It is because the company is severely in debt or has excessively high operating costs? The repercussions of 9-11 left the entire airline industry in retrenchment mode. One area hit hard was information technology: Delta trimmed 250 IT positions, Northwest some 300 positions, and United another 700 workers.[15] No matter what causes a business crisis, the key to solving it is for management to understand the cause and be ready to address it realistically. After understanding the cause, there are many different actions a company can take to try to achieve a successful business turnaround. Some of these actions can include selling off assets, pruning, cutting costs, boosting revenues, revising strategies and a combination of efforts.

Selling off assets In most cases of business crisis, cash flow is a serious concern. One of the quickest ways to generate cash is to sell off some of the company's assets. This could include plant and equipment, land, inventories, or even patents.

Pruning One common action used in retrenchment is pruning. Pruning means either selling off or closing less profitable or older facilities, curtailing production of marginal products, cutting out company perks (such as company cars) and cutting back on things such as advertising, public relations and even customer service. One of the most common types of pruning is also one of the most uncomfortable: reducing the workforce or implementing layoffs.

Cutting costs Cost cutting works when the firm's value chain and cost structure can allow for some rather serious radical surgery. Sometimes part of the problem is operating inefficiencies. This might be the result of operating costs that are out of

---

[14] Tom Fitzgerald, "Corporate Retrenchment or Corporate Renewal?," *The CEO Refresher*, 2001, <http://www.refresher.com/!corprenewal.html>.
[15] D. Lewis and M. Koller, "Airline IT Upgrades, Staff Hit Hard," *InternetWeek*, 11, 2001, <http://www.internetweek.com/showArticle.jhtml?articleID=6402648>.

line. It could mean that too much is being spent on component parts or that the value chain includes a number of low-value adding steps.

Boosting revenues Boosting revenues is an action that can be taken when there is little or no room in the operating budget to cut costs. Boosting revenues can include cutting prices, increasing advertising, bolstering the sales force and embarking on quick and easy product improvements. Boosting revenues can be very successful at getting results if the customer is relatively price sensitive.

Strategy revision Sometimes weak performance can be caused by a bad strategy. There are many options a company can take with regard to strategy revision. It might mean merging with another firm (an example of this would be the acquisition of KLM by Air France), shifting to a new competitive approach, or retrenching to the core products of the company. When Sears went through a retrenchment during the late 1980s, they chose to concentrate on the core products on which the company was built nearly 100 years previously.

Combination As with any strategy we have discussed, sometimes the best alternative is a combination of several different actions. The tougher the problem, the more difficult it will be to turn the company around and the more likely one approach may not be enough. In many cases a turnaround manager is brought into the picture. What is a turnaround manager? It is a person whose sole job is to save the company from financial ruin. Turnaround managers are often given free reign to do whatever they see fit to get the company back on target. With or without a turnaround manager, turnaround efforts tend to be very difficult, involving high risk and often fail.

US Airways' Woes[16]

In August 2002, US Airways was the first major airline in the United States to file for bankruptcy protection since September 11, 2001. US Airways had the highest seat-mile cost in the industry. In retrenchment mode, US Airways had to submit a reorganization plan with enough profit to meet a repayment schedule. Early in its plan, it cut thousands of workers, including mechanics and related workers, fleet-service workers, customer service agents, pilots and flight attendants – and pilots saw a 33 percent cut in pay. It also shut down the Florida maintenance and reservation station and sought concessions from existing workers. Six months into the retrenchment effort the company had already cut $1.3 billion in annual costs.

---

[16] Laura Williams-Tracy, "US Airways Needs Hub Here for a Turnaround," *Charlotte Business Journal*, 6 January, 2003, <http://www.bizjournals.com/charlotte/stories/2003/01/06/focus2.html?page=2>.

What are the Common Themes to Turnaround?

Anthony Henfrey of the Postern Executive Group, a specialist in business turnaround, identified six common themes of successful business turnaround.[17]

- First, a company that is likely to turn around successfully is one that has some genuine potential. The potential could be brand name or market leadership, but it is something upon which the company can depend.
- Second, one of the reasons that businesses fail is because many of them are a conglomeration of unrelated businesses, possibly the result of previous acquisitions. What is needed is some kind of clarity, returning to the company's core strengths. With this clarity, turnaround is more likely to be positive.
- Third, a struggling company generally needs to have an outsider's perspective, which may be more objective. A turnaround manager can generally see things to which the company management is either blind or chooses not to see.
- Fourth, the company does not need to forget about its experienced managers. Successful turnaround efforts still need experienced managers to run the business.
- Fifth, faltering companies often suffer from fuzzy lines of responsibilities. For example, sometimes it is unclear who is really in charge. This needs to be cleared up if the company has any hope of survival.
- Sixth, increasing inventory turnover is often not the best solution. If a company is going to turnaround, succeed and grow, it first needs to concentrate on making a good product rather than simply selling off inventory.

Turnaround is only successful if the company is growing again, but, according to Henfrey, the secret behind both successful and non-successful turnaround efforts is the quality and the actions of management.

Liquidation

Liquidation is often considered the strategy of last resort. Sometimes the business is just too far gone to be turned around and saved. In a situation like this, it makes the most sense to close down the business and sell off its assets, which is what is meant by liquidation. Liquidation results in the loss of jobs in addition to the loss of a business. It is considered the most difficult of strategies because it affects not only the general community inside the business, but also the community in which the organization resides. Early liquidation is best, because prolonging it only results in more costs and more ill will.

---

[17] A. van de Vliet, "Back from the Brink," *Management Today*, January 1998: 36–40.

Divestiture

Divestiture refers to selling off business components. Some companies treat a divestiture as a painful event that they never want to revisit. Actually, divestiture is often another strategic choice for a diversified company. Highly diversified companies often find it is very difficult to manage such businesses and decide that they are better served if they can narrow their diversified base and concentrate on a more limited scope of operations. According to a 2002 KPMG study, the three key strategic objectives for divesting are the following:[18]

- Increasing Shareholder Value .... 84 percent of respondents.
- Improving Profitability .... 73 percent of respondents.
- Emphasizing Core Competencies .... 71 percent of respondents.

Once the decision is made to divest, the study showed that the tactical goals of divestiture are to maximize the purchasing price and maintain the support from management and the board.

Many companies see divestiture as a natural occurrence in the life cycle of a business and feel that divestiture can be part of a forward-thinking corporate strategy rather than a painful event. Recent thinking about divestiture is that "divestiture is not an admission of failure but an important way for a company to succeed."[19]

Methods of divestiture There are two basic means of divestiture: selling the business outright or spinning the business off as a financially independent company. The latter makes sense when the company to be spun off has some long-term profit potential.

Selling off the business sometimes gives rise to many different concerns. For example, when Irish air carrier Aer Lingus planned to sell its aircraft maintenance subsidiary team to FLS Aerospace, the biggest problem was to convince the workforce of the deal's profitability in order to get the sale approved. Obtaining approval became more of a problem than initially believed.[20]

Sometimes trying to find the appropriate buyer can be the most difficult part of selling off a business. The company must consider not only trying to sell the business but also focus its attention on what kind of buyer would benefit the most. In other words, where would the business best fit? If a suitable business cannot be located, management must then decide whether to spin off the business, in the event the business still seems viable, or to liquidate its assets.

---

[18] Kevin McDermott, "Divestitures a Solid Shareholder Value Option in Tough Times," *Nashville Business Journal*, 30 May 2003, <http://www.bizjournals.com/nashville/stories/2003/06/02/smallb6.html>.

[19] Ibid.

[20] T. Gill, "AerLingus Is to Lose Team," *Airline Business*, 13, no. 3, 1998: 10.

Setting Corporate Direction

> ### Dobbs International Sold to SwissAir Parent[21]
>
> In 1999, Viad Corporation's food service business divested its Memphis, Tennessee-based airline caterer, Dobbs International, to SAir Group Holdings, the parent corporation of SwissAir and Gate Gourmet Co. Gate Gourmet was a $1.1 billion airline caterer, operating in Europe, Asia and Latin America. The combination of the two units made Gate Gourmet the largest airline caterer in the world.
>
> The sale by Viad was aimed at paring down some of the company's corporate debt and opening up opportunities to make other acquisitions more valuable to their core business. Viad spokespersons said the move was consistent with their long-term strategic plan to reallocate assets into high-growth companies with margins that would meet their corporate goals.

International Strategies

There comes a point in the life cycle of an industry when the domestic market does not seem to offer much hope for sustainable competitive growth. As previously discussed, some companies choose to embark on vertical or horizontal integration to try to remain viable and sustain profitability. Another option is to engage in a multinational strategy. International diversification provides the company an opportunity to drive down costs by expanding to other countries. In other words, it allows the company to take full advantage of economies of scale. It also helps the company to achieve a greater return on R&D costs. Specifically, in an environment where technological obsolescence is great, investing in new technology can be almost cost prohibitive for a company operating in a purely domestic market. If, however, the length of time to recoup this expenditure can be expanded by entering into an international market, it is easier for the company to recover from its R&D investment.[22]

International diversification, however, comes with its own set of risks. Trade barriers, logistical costs, access to raw material, distance from the home office and cultural differences can cause tremendous problems to a company trying to implement an international strategy.[23] Still another problem often associated with international

---

21 P. King, "Viad Corp. Sells Dobbs International to SwissAir Parent," *Nation's Restaurant News*, 33, 1999: 18.
22 M. Kotabe, "The Relationship Between Off-Shore Sourcing and Innovativeness of US Multinational Firms: An Empirical Investigation," *Journal of International Business Studies*, 21, 1990: 623–638.
23 M.A. Hitt, R.E. Hoskisson and H. Kim, "International Diversification: Effects on Innovation and Firm Performance in Product-Diversified Firms," *Academy of Management Journal*, 21, 1997: 767–798.

diversification is that of economic risks. Primary among these risks are differences and fluctuations in currencies. The value of different currencies can have a major effect on a company's competitiveness on the global front because it affects the price of goods produced in different countries. For example, if the value of the dollar increases, it can be detrimental to U.S. firms exporting to international markets because of the product price differential.[24]

In the airline industry, regulatory constraints have made international diversification almost prohibitive. Most countries have caps on foreign investments in the airline industry where the issue of airline ownership and effective control becomes problematic. VirginBlue in Australia is an exception to the typical practice of ownership restrictions, as Australia is one of the few countries with a freer regulatory environment than most countries, both in terms of ownership rules and air services.

Strategies Tailored to Specific Situations

Strategies for Fragmented Industries

What is a fragmented industry? A fragmented industry is one that is characterized by no one industry leader but, instead, many small to medium-sized firms. No one company has the market share and no one company can influence the industry outcome.[25] The airline industry is an example of a fragmented industry.

Characteristics of Fragmented Industries

- Market demand is extensive and diverse and, therefore, a large number of firms can co-exist.
- Low-entry barriers allow small firms to enter quickly.
- There is no opportunity for scale economies and, therefore, small companies can compete equally.

The small size of most companies in fragmented industries puts the company in a relatively weak position to bargain with suppliers and buyers. However, companies can cultivate a loyal customer base and grow faster than others in the industry. How?

---

[24] M.A. Hitt, R. Duane Ireland and R.E. Hoskisson, *Strategic Management Competitiveness and* Globalization, 4d edn, Cincinnati: South Western College Publishing, 2001, 349–350.

[25] Marc Dollinger, "The Evolution of Collective Strategies in Fragmented Industries," *Academy of Management Review*, 15, no. 2, 1990: 266–285.

Some do it through special programs, like Delta's SkyMiles, American's Advantage, and other frequent flier reward programs.

Some of the most popular competitive strategies for companies in fragmented industries include the following: [26]

- *Low-cost operators* Low overhead, low labor costs, high productivity, and dedicated operational efficiencies. This is a typical strategy undertaken by several low-cost carriers. They have achieved it through outsourcing of services (e.g. maintenance, training, etc.), leasing aircraft rather than buying them, providing monetary or other incentives for higher productivity and focusing on very high aircraft utilization.
- *Specialization by product type* When the products of a fragmented industry include a range of styles and services, a company can choose a strategy to focus on one product or service category. This has also been a typical strategy by most LCCs that focused on discounted price transportation in a single (economy) class configuration with "no-frills" provided to their customers.
- *Specialization by customer type* A firm in a fragmented industry can stake out a niche in the marketplace by catering to a special type of customer. Air Ambulance service providers do just this as do charter operators. Also, several airlines, for example Lufthansa, focus on business-class only flights on specific city pairs.
- *Focusing on a limited geographical area* A firm in a fragmented industry can concentrate on a local or regional geographical area. Several airlines have this scope of operations. An example is Commutair, which is based in Plattsburgh, New York and serves 19 cities in New England.

Strategies for Declining Markets[27]

What happens if the industry that a company exists in is declining? Some might think that the best strategy for a declining market is to get out as fast as possible. In some cases, this is the best strategy. Yet there are other options that can be taken where a company might end up owning the majority share of the market. If the competitor in the industry has dim prospects for the future, it probably makes sense to cut back the business or cut all losses. However, it is important to realize that stagnant demand is not enough to make an industry unattractive. Sometimes selling out is not the best idea. What should a company do?

- Pursue a focused strategy that is directed toward the fastest growing segment of the market. Aerospace manufacturers have provided us with examples of planning

---

[26] Michael E. Porter, *Competitive Strategy*, New York: The Free Press, 1980, 201–202.
[27] Porter, 254–274.

based on assumptions from their fastest growth market segments. Interestingly, different manufacturers have interpreted the growth segments in different ways. These results, in some cases, have varied significantly and as such they have produced radically different strategies. An example is the decision of Airbus to focus its attention on the A380 while Boeing focused on the Boeing 787. Airbus chose a very large aircraft to be the driving force behind its future strategy, whereas Boeing chose a radically redesigned aircraft of medium capacity with superior economics and innovative design.
- Stress differentiation, but differentiation based on innovation. Differentiation without being tied to new innovations will likely do nothing to save the industry. This concept is tied to the practice of entrepreneurship.
- Try to drive costs down and become the low-cost producer in the industry.

Strategies for Emerging Industries

Organizational life cycle theory suggests that companies go through various stages similar to the stages of human development. Industries also experience these growth cycles. Thus, when considering strategies for competitive advantage, it is important to keep in mind the point on the industry life cycle. For example, one would not choose the same strategies for firms in emerging industries as he or she would for firms in declining industries. If a company is active in an emerging industry, there is a lot of guess work in what will or will not work. Everyone in the industry is a pioneer and will be trying to outsmart the competition.

First-Mover Advantages

An important strategic choice for competing in an emerging industry is determining the appropriate time of entry. Porter suggested several general circumstances when early entry is warranted, which are listed as follows:

- If the image or reputation of the company is important to the buyer and the firm can enhance its reputation by being the first on the market.
- If it will be difficult for followers to imitate the technology, then the company will not be hurt by successive technological generations of the product.
- If customer loyalty will be enhanced by being first to the market.
- If an absolute cost advantage can be realized by being first to the market.[28]

The disadvantages, however, come with the fact that being the first mover is also risky. As noted by Porter, "early entry (or pioneering) involves high risk but may involve

---

[28] Porter, 232–233.

otherwise low barriers and can offer a large return."[29] The first mover is, in essence, working out the "bugs." As soon as they do this, in some cases, their strategy can be copied. On the other hand, sitting and waiting is not always a bad idea. There are times when what the first mover did is very easy to imitate and can be improved and perfected upon. The follower has an opportunity to perfect the more primitive technology developed by the first mover. As a result, sometimes the follower can progress into the next generation of products leaving the leader behind. The decision on whether to be a first mover or a follower is generally determined by how much risk the company is willing to take.

Case Illustration: AA's Acquisition of TWA: Timeline of Events[30]

With a January 10, 2001, filing at the Securities Exchange Commission (SEC), American's bid to acquire TWA became official. TWA announced its plans to enter Chapter 11 bankruptcy protection and accept a buyout by AMR Corporation's American Airlines. American offered to pay TWA $500 million and assume about $3 billion of aircraft operating leases. On January 25, 2001, however, Continental Airlines objected to American Airlines' buyout of TWA and indicated that it was prepared to spend up to $400 million for some assets of TWA in a deal that would let TWA continue flying as an independent airline. Northwest Airlines also objected to American Airlines' plan to buy most of TWA's assets. Consequently, a TWA spokeswoman said the airline was "prepared to look at higher and better offers."[31]

Other interested parties also weighed in on AMR's offer. On February 17, 2001, TWA pilots submitted their official support of the proposed sale of TWA assets to AMR Corporation by notifying the U.S. Senate Committee on Commerce, Science and Transportation. The pilots, represented by the Air Line Pilots Association (ALPA), also said the American buyout provided job security and protected medical and retirement benefits for pilots and other TWA employees.

On March 1, 2001, AMR and Jet Acquisitions Group Inc. submitted official bids for Trans World Airlines. The AMR bid, as mentioned earlier, was for $500 million plus the assumption of $3 billion in debt and lease obligations, while Jet Acquisitions bid $889 million for "substantially" all of TWA's assets. On March 5, 2001, the auction of TWA's assets commenced and then recessed in light of an impending decision. TWA's board chose AMR's offer as the "highest and best" on March 7, 2001 and, five days later, the U.S. Bankruptcy Court entered an order approving the sale of TWA's assets to AMR. Final approval of the order came from the U.S. Department of Justice on

---

[29] Porter, 232.
[30] Flouris and Swidler, 173–180.
[31] Flouris and Swidler, 174.

March 16, 2001, and it set the stage for American Airline's announcement on April 9 that it had completed the acquisition of TWA's assets.

In the AMR 2000 Annual Report, the company rationalized its acquisition of TWA both in strategic and financial terms. Don Carty stated the value of TWA's acquisition to shareholders:

> While our 2000 network progress was impressive, three transactions announced in January 2001 represent a giant leap forward for our network building efforts. First, we agreed to purchase substantially all the assets of TWA for approximately $625 million in cash and the assumption of over $3 billion of TWA's obligations.[32]

Carty's reference to a second and third transaction included obtaining certain assets from U.S. Airways along with leasing gates and slots from United Airlines and acquiring interest in a new start-up airline, DC Air. However, the consummation of the DC Air transaction, as well as American's acquisition of assets from U.S. Airways, was contingent on the closing of United's proposed merger with U.S. Airways. When the merger between United and U.S. Airways was officially called off on July 26, 2001, the last two points of AMR's strategic plan were left unfulfilled. This marked the first in a series of events that would alter the airline industry's competitive landscape and ultimately the financial stability of AMR.

On September 11, 2001, four terrorist plane crashes caused a shut down of all commercial airline traffic in the U.S. for several days. To guard against future attacks, the federal government required new security measures before flights could resume. Nevertheless, customer demand fell significantly and several of the airlines reduced flight schedules by 20 percent or more. Despite some cost savings from smaller operations, airline losses began to mount, and the promise of government guaranteed loans did not stem the flow of red ink. Eventually, United Airlines and U.S. Airways declared bankruptcy to shield their assets from creditors.

At American Airlines, the tragedy of 9-11 meant that demand for its seats were slashed just months after increasing capacity and operating expenses with the acquisition of TWA. The fallout from 9-11 on the recently expanded American Airlines played itself out over several months. Documentation for a number of these important corporate events may be found in various Quarterly Reports (Form 10-Q) and Current Reports (Form 8-K) filed with the Securities and Exchange Commission. Following a second quarter loss in 2002 of $495 million ($3.19/share), Carty cited a "weak revenue environment" and, in response, the company stated it had cut capacity 10.4 percent from the previous year. On August 13, 2002, one month after releasing the second quarter results, American unveiled its next series of fundamental business changes. The company planned to reduce capacity another 9 percent, retire its fleet of Fokker

---

[32] Ibid.

100s and nine of its TWA 767-300 aircraft, eliminate 7,000 jobs, and delay delivery of 35 new planes. Only nine days later, American downwardly revised its capacity expectations for the remainder of 2002 and the entire year forecast for 2003.

Following new financial practices issued by the Financial Accounting Standards Board (FASB), American also announced in their 2002 second quarter report that it had completed the first step of impairment analysis. AMR found that its net book value exceeded fair market value (reflecting market price of securities) and it would likely record a pre-tax charge of up to $1.4 billion to write-off the company's goodwill. This was the first glimpse of the airline's estimate of diminished value due to the events of 9-11. On October 4, 2002, American confirmed the $1.4 billion charge and stated that the amount "related to the acquisitions of Reno Air, Inc., ACI Holdings, Inc. (AirCal), American Eagle acquisitions and certain assets from Trans World Airlines, Inc.."

Losses at American continued to increase through the year, and on January 22, 2003, AMR reported a 2002 fourth quarter net loss of $529 million ($3.39/share). "Clearly, results such as the ones we reported today are unsustainable," stated Carty. He went on to say that the company would seek to lower its labor costs and would also record a significant pension liability to be charged against equity. After negotiating with its three major unions, AMR announced on April 16, 2003, that it had reached agreements with their employees that would result in $1.8 billion in annual savings. The cost savings included $660 million from pilots, $340 million from flight attendants, $620 million from the flight mechanics union and $100 million from management and staff. Looking forward, the company would further attempt to renegotiate more than $800 million of debt to remain in compliance with bond covenants.

While American had been saved from filing for bankruptcy, its financial position remained weak and future prospects uncertain. In an April 23, 2003, announcement of first quarter results, Carty said, "We are beset on all sides by a struggling economy, the continued uncertainties regarding hostilities in the Middle East, concerns regarding the SARS outbreak, fuel prices that are significantly higher than they were a year ago and fare levels that are at 30-year lows." This would be one of Carty's last public statements, as word of a secret bonus plan for management surfaced following the announcement of the labor union agreements. One day later, the AMR Board of Directors accepted Carty's resignation and appointed Gerard Arpey as the company's new CEO. Thus, a turbulent chapter in American Airline's history ended following a series of events that included the acquisition of TWA, the tragedy and fallout of 9-11, a set of new labor agreements to save the company from bankruptcy and, finally, the resignation of its CEO and chairman.

Table 6.1 summarizes several salient facts from its 10-K reports filed with the Securities and Exchange Commission concerning operations for the years 2000–2002.[33] First, total market value of equity (number of shares outstanding times market

---

[33] Securities and Exchange Commission, American Airlines', *Form 10–K405 Annual Report*, 2000, <http://www.sec.gov/Archives/edgar/data/4515/0000950134-00-002448.txt>.

price) dropped from more than $7 billion prior to the merger to $1.2 billion following the events of the takeover and 9-11. Over this same period, earnings per share went from a positive $5.03 to a negative $22.57.

The drop in equity value and flow of red ink stemmed, in part, from American's substantially increased debt load. After assuming $3 billion worth of TWA's liabilities, American's long term debt more than doubled, increasing from $4.1 billion to $8.3 billion between 2000 and 2001. Capital needs further pushed long-term debt to almost $10.9 billion by the end of 2002. Fixed payments on a number of leased planes added to American's financial obligations. While American's fleet stood at 978 aircraft at the end of 2000, only 616 planes were owned. With the takeover of TWA, American's lease obligations increased significantly (operating leases increased from 240 to 383 aircraft). Moreover, many of TWA's planes were not put into operation by AMR either because of old age or difference in aircraft type from American's existing fleet.

The increased long-term debt load and lease obligations came just months before the tragedy of 9-11 and the resulting decline in demand for flying. Despite a fleet more than 20 percent larger than before the merger, passenger revenue miles increased only 4.4 percent between 2000 and 2002. With little change in its cash position, American Airlines could not generate the necessary revenue to pay its fixed obligations and found itself on the brink of bankruptcy in early 2003.

Table 6.1 American Airlines Corporate Information Pre- and Post-TWA-Takeover[a]

	2000	2001[b]	2002
Market Value of Equity (millions)	$7,144	$4,065	$1,204
Earnings Per Share	$5.03	$–11.43	$–22.57
Long Term Debt (millions)	$4,151	$8,310	$10,888
Current Liabilities (millions)	$6,990	$7,512	$7,240
Cash on Hand (millions)	$89	$120	$104
Owned and Leased Aircraft Operated by AMR			
Owned	616	679[c]	681
Capital Lease	122	125	109
Operating Lease	240	353	315
Owned and Leased Aircraft Not Operated by AMR			
Owned		36	21
Capital Lease		6	16
Operating Lease		30	37
Total Number of Aircraft	978	1229	1179[d]
Revenue Passenger Miles (millions)	116594	120606	121747

[a] Acquisition of TWA assets completed on April 9, 2001. All figures include American Eagle assets and liabilities.
[b] Includes data after acquisition of TWA's assets and liabilities.
[c] Of the 679 owned planes, 19 were formerly TWA aircraft.
[d] Total includes 27 Boeing 757s and 103 MD-80 aircraft that were formerly TWA planes.

Chapter 7

Establishing a Strategy

> Success is making money, not in the size of the airline.
>
> Gordon Bethune[1]

A good strategy most come from a holistic view of the business, taking into account both the external pressures as well as the real issues surrounding the capabilities of the firm. The tool most commonly used to begin this holistic assessment is a SWOT analysis.

SWOT Analysis

Determining a company's internal strengths and weakness and external opportunities and threats, SWOT analysis helps to provide a good overview of where the company stands competitively. The premise of the SWOT analysis is that a good strategy must be the byproduct of the company's internal resources and capabilities and the situation in the industry.

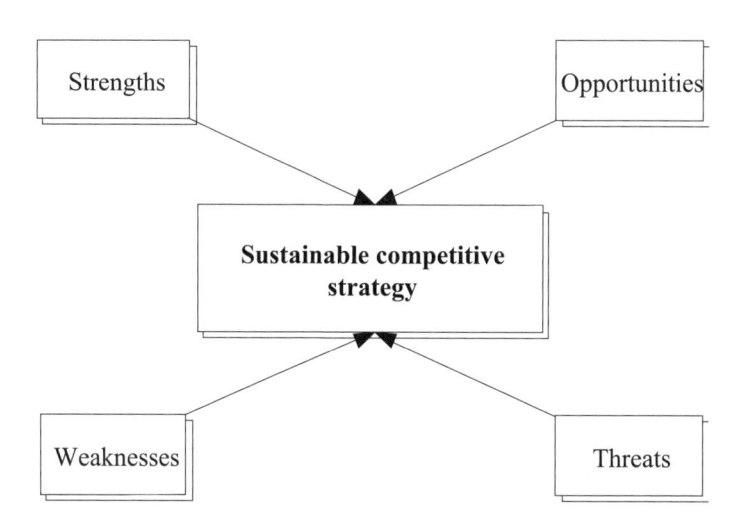

Figure 7.1   SWOT Analysis

---

[1]   See <http://www.skygod.com/quotes/airline.html>.

Strengths

What can be considered a company's strength? Identifying company strengths should be easy if one does a thorough analysis of the internal environment. A company's strength is something the company simply does well, or something that is a positive asset. Think of a strength as something over which a company has control. For example, the company has control over its core competencies in establishing and implementing them. It also has control over its resource capabilities as well as its corporate balance sheet. A strength is always going to be internal to the company. What are the sources of strengths? Strengths can easily be found by examining the company's value chain. Another way to find strengths is through an examination of the functional areas within the company.

Sources of Strengths

- The employees and their expertise.
- A strong financial position.
- A strong brand name.
- Brand loyalty.
- Quality product.
- Strong knowledge management.
- International operations.
- Well-oiled operating procedures.
- Good supplier or customer-relations.
- Strong promotional practices.

Keep in mind that the company's strengths are those factors over which the company has control. Strengths are found through a thorough analysis of the internal environment and a close examination of the company value chain.

Weaknesses

Like strengths, identifying company weaknesses should be easy if one performs a thorough analysis of the internal environment. A weakness is something that the company does not do well and over which it has control. For example, the company has control over its financial picture, but this does not always mean that the money is being managed appropriately. It also has control over its products, but this does not mean that the product is good. Maybe the product is no more than a copy of a competitor's product. Will this give the company a competitive advantage? Not likely.

A weakness is internal to the company. What are the sources of weaknesses? Weaknesses, like strengths, can be found by examining the company's value chain or by taking a close look at each of the functional areas. The most important thing about determining weaknesses is to be honest. It is always easy to determine strengths because these are the things of which the company is proud. But does management generally want to admit weaknesses? Probably not. Strategic analysis is no better than the time and effort put into the analysis and no better than the honesty of the evaluation. Therefore, it is essential in the analysis to admit to all weaknesses. Also, keep in mind that something can be a strength and a weakness at the same time. For example, a company might have a very motivated workforce (strength), but they might not have the expertise needed to move the company into the next generation product (weakness). In this scenario, the employees can be considered as both strengths and weaknesses.

Sources of Weakness

- Old, rundown facilities.
- Lack of any computer integration.
- Unused capacity.
- High inventory costs.
- Large amount of obsolete inventory.
- No strategic direction.
- Sub-par product quality.
- Lack of good research and development.
- Lack of strong leadership.
- Lack of corporate vision.
- No product recognition.

Opportunities

The next area that must be addressed to complete the SWOT analysis is the external analysis. After a thorough analysis of the external environment, company management should be able to determine the opportunities and constraints that face their industry. These are things within the environment that the company has absolutely no control over. And, as previously discussed, there are different levels of these opportunities and threats. Some exist within the industry, but others, like economic issues, exist at the macro level. Nevertheless, these are things that have an effect on the way a company conducts business.

What are sources of opportunities? One should examine the external analysis. If a thorough assessment of the external environment was conducted, these opportunities should be easy to spot.

Sources of Opportunities

- Untapped international markets.
- Untapped customer needs.
- Large demand for the products produced by the company.
- Favorable demographic conditions.
- Acquisitions of rival firms or firms in similar industries.
- Acquisition of firms that will facilitate backward or forward integration.
- Changes in international trade policies.

As evident above, opportunities are factors that exist in the external environment of which the company can take advantage. Company management cannot control opportunities but they can certainly utilize an opportunity to further the company's strategic position.

Sources of Threats

- High unemployment.
- High periods of inflation.
- International competition infringing on the market.
- New government regulations.
- Bargaining power of the customer has heightened.
- Supplier has increased prices.
- Substitute products.
- New domestic competition.
- Demographic changes.
- New innovations that render the existing product obsolete.

Threats

A company has no control over a constraint or threat, but the threat can do a lot of damage to the business if not managed properly.

When the airline industry in the U.S. was deregulated, this might have been thought of as both an opportunity and a threat. The opportunity came from the fact that the airlines were now in charge of their own destiny, for instance setting their own routes and prices, to name a few. The threat came in the fact that the airlines were now in charge of their own destiny, and they had never done this before! The cushion provided by the government was no longer there.

Jet Airways, in India, provides an applied example of the SWOT analysis methodology in an airline industry setting. Table 7.1 provides a SWOT analysis of Jet Airways.

Jet Airways was incorporated on April 1, 1992, and commenced its operations with four B737-300s on May 5, 1993. At the time (until October 1997), the airline had 20 percent participation from Gulf Air and Kuwait Airways, each, with the remaining balance held by Tailwinds Ltd., an OCB company registered in the Isle of Man and owned entirely by Naresh Goyal. Implementing a new regulatory framework effective October 1997, the government of India withdrew permission for foreign airlines to invest in Indian domestic airlines, which, in effect, led to Tailwinds Ltd. acquiring the equity stake previously held by the two Middle Eastern carriers. Jet Airways went public in March 2005 and posted a profit, up 103 percent from the previous year, of $89.4 million USD. Revenue for the year 2004 was $1 billion USD, an increase of 24 percent from the previous year.[2]

The key driver for the phenomenal success achieved by Jet Airways was its well-defined and clear-cut mission statement. The philosophy of Jet Airways' mission statement was driven by customer service, which was adopted to ensure that the airline became the "World's Best Domestic Airline" by providing an unmatched quality and standard of service and an extremely high degree of reliability of operations. Yet the fundamental objective from the beginning was to recognize the quality of service and acknowledge that the rationale of the airline's existence and success would be customer acceptance and loyalty. In addition, the importance of teamwork and coordinated efforts towards building the most preferred airline in India were recognized as key factors for the overall success of the airline.

Since its inception, the airline has grown into a formidable contender in the Indian domestic market. In order to determine Jet Airways' future prospects, it is essential to understand the sources of its past success. It is equally as important to understand the relevance of these achievements in today's market in an effort to ensure continuing success. In this context, a SWOT analysis of the airline can be very beneficial.

After identifying the components of the SWOT analysis – strengths, weaknesses, opportunities, and threats – the next step is to get from the SWOT analysis to developing competitive strategies. The strategies should be a marriage between the internal environment and the external environment and should provide a natural fit with the company's internal resources. In other words, management should use strengths in the internal environment to either exploit some identified opportunities or to protect against external constraints. Sometimes the company can be turned around with the right opportunity. However, at other times, it makes more sense to close down or sell out the business because the weaknesses are so great and constraints unmanageable. This is what has happened in the aviation and airlines industries. For example, as

---

[2] *Commercial Aviation Today*, 19 May 2005. See http://www.commercialaviationtoday.com/archives/2005/3-3087-cat-19-May-05.htm>.

Table 7.1 Jet Airways' SWOT Analysis[3]

Strengths
- Emphasis on customer service and customer service relationships.
- Better passenger services compared to the competition, especially Indian Airlines.
- High aircraft utilization: the best in the Indian airline industry for the B737s and very high for the ATR72s.
- Youngest fleet age 3.14 years (2003) as compared to Indian Airlines' 14 years and Air Sahara's 5 years.
- Lowest number of employees per aircraft in India (171 per aircraft), despite limited out-sourcing opportunities in the country.
- All India network of Jet Air offices (110 offices), a General Service Agent (GSA) company established by Naresh Goyal in the 1970s.
- Strong focus on cost leadership and benchmarking.

Weaknesses
- Too much dependence on the business travel market segment.
- Increased dependence on passenger revenues rather than having a diversified source.
- Domestic airline with no exposure in the regional-international segment.
- Limited viable expansion possible into newer destinations over and above currently served.

Opportunities
- Alliances, joint ventures and the establishment of international cargo warehouses, should help improve revenues than what is being achieved by the passenger revenue segment.
- Further liberalization of the Indian aviation sector will help the airline with exposure to the international markets in order to compete effectively with other international airlines.
- Given the growth of key destinations, it is possible for the airline to further increase its market share by increasing capacities on the routes.

Threats
- Air Sahara, Kingfisher, other emerging low-fare airlines and a renewed Indian Airlines, may attempt to erode the existing market share held by the airline. Its competition airlines may erode the share by waging a fare war campaign which may lead to a shift of price-sensitive travel segment.
- Possibility of new niche players eroding market share on the regional routes.
- The dominant position held by the airline is difficult to be sustained without ensuring that the airline does not fall into complacency.

---

[3] Gilbert George (Manager Jet Airways), interview by Triant Flouris, Montreal, Canada, August 2003.

a result of the Deregulation Act of 1978 in the U.S., the competitive landscape in the U.S. airline industry changed dramatically. Several carriers, such as Pan Am and Eastern, were not able to change their business models effectively to maintain a viable competitive position in the deregulated environment. Today, regulatory realities, deregulation, and liberalization of the competitive landscape remains a very important issue. In 2003, the ICAO Air Transport Conference (ATC/5) focused not only on whether to liberalize the market but on how to do so. At the end of this chapter the issues surrounding liberalization and how these issues may be playing out in the future are discussed in depth.[4]

In the context of liberalization or deregulation, Eastern and Pan Am can be cited as examples of failed ventures while Southwest can be seen as a success. Without a deregulated business environment, which gives an airline the right to determine its capacity and decide on the price it should charge, Southwest Airlines (the low-cost business model) would be nonexistent. On the other hand, Pan Am and Eastern, two airlines that had significant internal strengths operating under an economically regulated regime, found deregulation to be an impenetrable barrier. The technically competent airlines, which thrived in an era of regulation, were not able to adapt and accommodate deregulation and, as such, both companies shut down.

Organizational flexibility is key to transforming an internal weakness into a strength or an external threat into an opportunity. Currently, airlines that are well suited to capitalize on opportunities, which may exist as a result of further liberalization internationally, are those that have an organizational structure that is flexible enough to provide customers with what they need in the future. Air transportation, today, is not a luxury commodity as it was in the 1950s and 1960s. Operators who can address the needs of their core customers via distinctive competences, whether they are leisure or business customers, will be the ones who succeed.

If a business has been able to establish a distinctive competence, its strategy must support this competence. However, a distinctive competence will not last forever unless it is cultivated. In order to cultivate a competence, one must look toward the future and make sure that management continues to build the strong resource base that got the company to its present state. One should look to next-generation technology or perfect operational processes so that, as discussed previously, it will be next to impossible for the competition to reproduce.

One thing to be aware of is that strategy does not always mean growth. Sometimes the best thing a company can do is to close down and cut its losses. At other times, it makes more sense to go through retrenchment (as discussed in Chapter 6) until the business is strong enough to move forward.

Another important point is that, with any given strategy, often there are critical issues that must be addressed prior to moving forward. Generally, these critical issues

---

[4] Adrianus D. Groenewege, *Compendium of International Civil Aviation*, 3d ed., Montreal, Quebec: IADC, 2003, 97–99.

are internal to the company, like an unfavorable balance sheet or poor, inefficient equipment. It may be imperative for management to clean up the internal mess before moving forward with a new strategy. If it is a cash flow problem, then maybe it is time to sell off some unproductive assets. Overall, moving forward with a new strategy may not be possible until some of these critical elements are handled.

What else Can be Gained from SWOT?

The SWOT analysis can give management a good snapshot as to where the company is at the present moment. It also helps to assess exactly where the company is in terms of the internal organization. Sometimes, through SWOT, it gives management the best overview from which to determine exactly what must be done in the near future to improve competitiveness. For example, SWOT might reveal to a company that there is great urgency to make changes in the organization if it wants to remain viable in the market. Thus, while SWOT helps to formulate strategies, it also gives a very thorough, and hopefully unbiased, overview of the company.

Formulating Strategy under Uncertainty

Wouldn't it be nice if the world were certain? Wouldn't it be nice if everything was transparent and whatever the constraints and opportunities identified in today's environment remained the same? Wouldn't it be nice to know that loyal customers will always remain loyal, no matter what the competition tries to do? Certainty would be wonderful, but it may also be a bit boring.

We live in a very uncertain world and formulating strategy in an unpredictable environment can be very trying at times. Yet while flux creates challenges, it also gives rise to opportunities. Although uncertainty brings the realization that some of the best determined strategies might not work out, this can be exciting and a source of opportunity. Most people are not fans of uncertainty, but stable, steady environments provide fewer opportunities in which a company can grow and prosper.

How does a manager prepare a company to take advantage of the opportunities that might come about as a result of uncertainty? They must first understand the drivers of change and then formulate strategic responses to deal with these drivers. Some suggest finding a strategy robust enough to weather any situation, although this is easier said than done. There is no one prescription for every situation, and if such a strategy exists, the risk verses payoff is likely to be watered down. Managers always want the highest possible payoff. If a strategy can fit a multitude of situations, is it really the best strategy for the company? Probably not.

E.O. Teisberg described six steps to developing insights into uncertainty.[5]

---

5   Elizabeth O. Teisberg, "Strategic Response to Uncertainty," *Harvard Business School*, #9–391–192, Harvard University Press, 1993, 1–12.

Step 1: Brainstorm a list of uncertainties that the firm faces and examine the logic of how these uncertainties may unfold Teisberg suggested making this list as expansive and as inclusive as possible, including possible changes in demand or supply, possible strategic moves or repositioning by the competitors, changes in the complexion of the competitors, and possible changes in technology, demographics, and substitute products. Further, she suggested including in the list any political uncertainties, economic, social, or even acts of nature. Finally, Teisberg noted that some of the outcomes are dependent on the actions of the competitors in the industry and that others are dependent on the firm itself. The key is to understand the interrelationships among uncertainties and between uncertainties and possible responses.

Step 2: Identify strategic choices the firm can make and consider the interrelationships among decisions and uncertainty At this point, Teisberg suggested developing a decision tree. Decision trees help a manager make a choice between several courses of action. They provide a structure within which the manager can lay out options and investigate the possible outcomes of the options. Decision trees also help to form a balanced picture of the risks and rewards associated with each option. Teisberg suggested that company managers ask themselves, within the structure of the decision tree and given the possible scenarios, what are the best options for the company? She cautioned not to look at decisions as good or bad or whether or not they constrain the strategies in the future. The reason for developing the decision tree early on in the process is that it provides a framework for getting ideas out on the table and looking at how they might unfold under different environmental conditions.

Step 3: Develop internally consistent visions of the future that challenge conventional views By developing scenarios it forces the strategist to look beyond simple forecasting and create some viable alternatives for the future. There is no set number of scenarios that should be developed, but it is important that the strategist develop enough scenarios to provide contrasting views. At this step it is important to ask such questions as the following:

1. What could possibly happen that would cause buyers to be more price-sensitive (or less price-sensitive)?
2. What must occur to change the situation in the industry such that profit margins plummet?
3. Are there possible substitutes on the market, or is there a situation where a product might be rendered obsolete?
4. Could a product become a commodity and, if so, what would this do to the industry?

There are countless other questions that should be addressed, but these are examples of those that might cause the strategist to think outside the box. While many people who support scenario development suggest having best and worst case scenarios, Teisberg said that it is more advisable to establish many different scenarios with credible outcomes.

Step 4: Check whether uncertainty is critical to decisions Sometimes it makes more sense to come up with several different alternatives, because one might find that uncertainty really does not have much affect on the strategic decision. In these cases, management does not have to give uncertainty another thought, at least for a while. If a firm's major decisions would be the same under every scenario, then uncertainty is a minor player and not worth the distraction. However, if a company's decisions would be the same under every scenario, it could also mean that the uncertainty was not characterized appropriately. Before going any further, a reassessment of the analysis should take place.

Step 5: Analyze how the firm's decisions may affect future uncertainties and the development of the industry Teisberg suggested that it does not make sense to assign probability to each scenario because it is not certain if the list of scenarios is complete. Instead, at this point, she suggested asking questions like the following:

1. What can the firm do that might influence some of the more important events?
2. Is there something that can be done to pre-empt moves by competitors?
3. Do regulatory outcomes depend on plans or commitments of the firms in the industry?
4. Is there a chance that some of the less aggressive competitors might become viable opponents?
5. Is there a way that management can reinvent the industry to avoid destructive competitive battles?

The last question is a particularly interesting one. Companies like Hanes redefined the hosiery industry by selling pantyhose in grocery stores. Maybe the firm can look at ways to reinvent the industry. An equivalent example from the airline industry is that of the advent of strategic alliances and the practice of codesharing. Airlines were able to expand their networks and marketing options via their participation in strategic alliances or through codesharing agreements with other carriers.

It is at this point in the analysis that management probably needs to analyze the decisions that were formulated in Step 2. The key is to use what was learned about uncertainty and determine the appropriate strategic choice to implement.

Step 6: Formulate strategic responses to uncertainty by considering the firm's activities throughout the value chain Remember, uncertainty is not likely to hit all aspects of the value chain at one time. Differing activities might be affected at different times. In other words, look at the whole picture. A manager wants to make sure that whatever choice is made supports the company's strategic direction; sometimes this means reducing the risk at some points of the value chain in order to gain the greatest payoff in the end. For example, maybe a company needs to stock more inventory than normal to be able to respond quicker to environmental situations.

Choices of Responses to Uncertainty

There is no one response that works best for all situations or all managers. Some managers are much more adverse to risk than others, and some companies have a history of conservative decision making. Managers have several categories of choices they can make:[6]

- Choices of Bets.
- Choices of Robustness.
- Spreading Risk.
- Hedging.
- Investing in Flexibility.
- Choices of Timing.
- Choices that Change Uncertainty.

Choices of bets Bets are commitments to high risk investments, and high risk usually brings about high payoffs. However, if management makes the wrong decision, they can also lose big! Of course, there is an element of "betting" in all management decisions, but good managers do not just bet. They make decisions based on some facts. This is often called taking risks (manageable) but not gambling (unmanageable). Managers need to look for the opportunities with the highest payoffs and at the same time be careful not to rely on only one plan. They must use other strategic responses to complement the bet.

Choices of robustness This means taking steps to reduce manageable risks. If management reduces risk across all activities in the value chain, then they are probably diluting the overall strategy and the likelihood of being competitive is drastically reduced. Instead, it is important to be selective about the choice of activities. How are manageable risks reduced?

Spreading risk The most common way to spread out risk is to invest in several different courses of action. For example, leading companies in the industry might want to be aware of all the new technologies on the market that could set new product standards. For example, managers in these leading companies will most likely want to protect the firm's leadership position. Therefore, to avoid being caught off guard

---

[6]   Teisberg, 1–12.

by a technology shift, they may choose to invest heavily in R&D as well as diversify R&D efforts. By spreading out risks in R&D, they can focus their investment further along the value chain in product improvements. Smaller firms, on the other hand, cannot afford to dilute their efforts by spreading resources over a wide range of products. There is always a drawback when choosing to spread out risk: the potential loss of strategic focus.

Hedging Another measure that managers might take to reduce risk is strategic hedging. Imagine a company that has developed new, next generation technology but is unsure how this will be received in the marketplace. Management is hedging its bet that this technology will be successful. However, to reduce the risk involved in failure, the company might also invest in the development of a less advanced, more common type of technology on which to fall back. Of course, the downside to this scenario is that management could end up spending more money and resources trying to avoid the risk than it might have cost to face the risk head on. Another example of hedging is exemplified in partnerships or joint ventures. Yet while managers may argue that these relationships provide a way for avoiding bad outcomes (e.g. competitive dominance by another company), these arrangements often mean that company knowledge, contacts and distribution channels are shared with the competitor. In this situation, the risk potential might surface some time in the future.

Investing in flexibility One last option for reducing manageable risk is through investing in flexibility. This could mean investing in new production facilities, which make it easier to change gears with relatively few switching costs. It could mean looking for ways to reduce the "brick and mortar" and change to a "point and click" environment. The idea of creating core competencies, as discussed in previous chapters, lends itself well to investing in activities and resources that create flexibility for future product development.[7] The key to investment in flexibility is to make sure the company outlook is looking toward the future. Management must always consider the question, by changing things now, will it pay off in the long run?

Choices of timing Strategic choices are not simple "yes" or "no" decisions. Timing can have drastic effects on the outcome of the strategy. Some believe timing is everything and in the case of strategic decisions it might be true. Knowing when to take a product to market or when to enter a new market can be imperative to the ultimate success of the strategy. Sometimes it is a matter of planning the strategy in multiple stages, testing the water before committing the company's resources and energy. Sometimes uncertainty makes people delay their decisions and delays could cause the strategy to fail. On the other hand, a delay may be exactly what is necessary for a strategy to

---

[7] Gary Hamel and C.K. Prahalad, "The Core Competence of the Corporation," *Harvard Business Review*, 68, no. 2, 1990: 79–91.

Establishing a Strategy

succeed. Sometimes delaying strategic choices allows the manager to understand the situation more clearly. However, some opportunities are only available for limited times. Strategists must learn to strike when the time is right!

Choices that change the uncertainty Sometimes management can change things to the point that they have changed the structure of the industry, or at least add new dimensions to the industry. How does one change an industry? They can negotiate new vertical or horizontal relationships; they can create barriers to entry; or they can tie up the distribution channels so that competitors must work harder to compete. Faced with the decrease in the purchase of CDs due to improved abilities to download and replicate music, managers in the music industry chose to change the industry structure by significantly reducing the retail price of CDs. How could they afford to do this? They had to reduce, simultaneously, the cost of producing the CD.[8]

As previously mentioned, one means of changing uncertainty is to change the basis of competition. Selling Hanes pantyhose in grocery stores rather than department stores was foreign to the industry. Likewise, when BIC introduced a disposable pen, this changed the industry structure. These are all important strategic examples of opportunities that ultimately had a profound effect on the given industries. Remember, the essence of strategy is dealing with the competition, finding the strategic point that will make the company successful.

Case Illustration: Background to Air Transport Regulation and Attempts for Liberalization[9]

In the light of the fundamental changes that have affected the air transport industry in the commercial and operating environment since the early 1980s, a Worldwide Air Transport Conference on the theme of International Air Transport Regulation: Present and Future was held in Montreal in 1994, under the auspices of the International Civil Aviation Organization (ICAO). The basic purpose of this Conference was to conduct an in-depth examination of issues, such as new regulations between States and market access for the airlines. Also discussed were criteria for licensing and investing in airlines, product distribution, and day-to-day operations of air carriers.

The Conference carried out the most extensive review of the bilateral system since the signing of the Chicago Convention in 1944 and formulated a set of new regulatory arrangements for future air transport regulation policies, to be used by States at their discretion for liberalizing their transport sector on the basis of the Chicago Convention. The outcome of the Conference in 1994 represented an important milestone in the evolution of the regulatory framework for international civil aviation. It was widely

---

[8] Teisberg, 1–12.
[9] Groenewege, 97–99.

recognized that the global air transport industry, as in other fields, is changing very rapidly and, as such, is challenging the ingenuity of civil aviation authorities, airlines, airports, aircraft and engine manufacturers and other providers of aviation products and services throughout the world.

Since the Fourth Worldwide Air Transport Conference in 1994, there have been significant changes in the regulatory and operating environment of international air transport brought about by economic development, globalization, liberalization and privatization, while ensuring the continued safe, secure and orderly growth of civil aviation worldwide. States have generally become more open and receptive toward regulatory reform, with many adjusting their policies and practices to meet the challenges of liberalization. In fact, considerable progress has been made with respect to air services arrangements at both bilateral and regional levels.

At the same time, the past decade also saw increasing involvement of some non-aviation organizations in air transport matters, such as the World Trade Organization (WTO) that came into being in January 1995, and along with it the entry into force of the General Agreement on Trade in Services (GATS). The GATS Annex on Air Transport Services applies trade rules and principles, such as most-favored nation and national treatment, to three specific air transport activities: a) aircraft repair and maintenance; b) selling and marketing of air transport services; and c) computer reservation system services.

Against this background, the Fifth ICAO Worldwide Air Transport Conference was held again in Montreal in March 2003, on the theme of Challenges and Opportunities of Liberalization. The Conference was preceded by a seminar in March during which prominent speakers from industry, government and academia discussed various strategic issues facing today's air transport industry and its regulators. The objectives of the Conference were to develop a framework for the progressive liberalization of international air transport, to include safeguards for fair competition, safety and security and measures for the effective and sustained participation of developing countries.

The Conference approved a Declaration of Global Principles specifying the individual and collective roles and responsibilities of States in working toward the ultimate goal of giving international air transport as much economic freedom as possible, in keeping with the needs of the traveling public and the industry, while respecting its specific characteristics and in particular the need to ensure high standards of safety, security and environmental protection.

In addition to the Declaration, the Conference adopted recommendations on air carrier ownership and control and on the role of ICAO in economic liberalization. The Conference also reached 67 conclusions, covering such areas as market access, fair competition and safeguards, consumer interests and product distribution dispute settlement and transparency. For a detailed overview of the action taken by the Air Transport Conferences of 1994 and 2003, as well as the conclusions reached and recommendations adopted by the two Conferences, see Regulation of Air Services under ICAO Air Transport Services.

Establishing a Strategy 139

At the IATA Annual General Meeting (AGM) and World Air Transport Summit (WATS), held in Washington DC, June 2003, IATA Member airlines called upon governments to help them ensure a sustainable future for international air transport. Accordingly, the Washington Declaration was adopted which outlines the following six measures needed to foster change and to ensure the long-term financial profitability of the air transport industry:

- Liberalization of ownership rules to allow airlines access to global capital markets.
- Economic regulation of airport and air traffic services (ATS) providers to eliminate monopoly practices.
- Increased cooperation and standardization between national competition authorities.
- Government restraint in imposing discriminatory taxes and fees on the aviation industry.
- Protecting consumer interests served by industry standardization from unilateral government measures.
- Governments assume their responsibility for security and the associated costs.

In February 1995, the U.S. and Canada signed a new Air Transport Agreement, reflecting the open skies concept; a very substantial liberalization of air services between the two countries without any limitations with respect to the volume of traffic, frequency or regularity of service, or the aircraft type or types operated by the designated airlines. The new agreement also acknowledges that market forces shall be the primary consideration in the establishment of prices for air transportation with minimum government regulation for the purpose of promoting trans-border commercial air services to the fullest possible extent. This brought about a rapid expansion of additional air services between Canada and the U.S.

In the course of June 1995, the U.S. also signed open skies air transport agreements with Austria, Denmark, Finland, Iceland, Luxembourg, Norway, Sweden and Switzerland, similar to the agreement signed with the Netherlands a few years earlier. These liberal bilateral agreements mark an important milestone in the development of international civil aviation by creating a free market for airline services between those countries. Each of these agreements gives the designated airlines of both countries the right to operate air services from any point in one country to any point in the other, as well as to and from third countries.

The agreements also give airlines pricing freedom and unlimited airline capacity, among other things. This development will have a significant impact on future bilateral air transport agreements to be negotiated between nations. In fact, it can be expected that an increasing number of open-skies air transport agreements will be signed between governments in the future. By mid-2002, about 85 open skies bilateral agreements had been concluded since 1992, involving approximately 70 countries, with the U.S. as one of the partners in two-third of the cases. These agreements involve not only developed countries but also an increasing number of developing countries.

Among the bilateral air service agreements concluded in 1996 was an agreement between Australia and New Zealand which provided for a single aviation market with the unrestricted right to fly anywhere in the other country. In the African Continent, States sought to increase cooperation by implementing the provisions of the Yamoussoukro Declaration of 1988 concerning liberalized traffic rights for African airlines, particularly at the sub-regional and regional levels. The Yamoussoukro Decision of 1999 declared open skies for African nations when the tourism sector was at its peak and was expected to grow. However, recent developments have indicated that the Yamoussoukro Decision of 1999 does not provide adequate protection to developing local airlines.

In June, 1996, the Council of Ministers of the European Union (EU) approved a two-phase plan for negotiating an air service agreement with the U.S., with the first phase including such topics as airline ownership limits, competition rules, code-sharing and computer reservation systems (CRSs); the second phase to include traffic rights and directly related aspects, such as capacity and tariffs. This with the understanding that existing bilateral agreements between EU States and the U.S. would be respected and movement to the second phase would have to be authorized by the Council of Ministers.

In recent years, significant progress has been made toward the creation of a Transatlantic Common Aviation Area (TCAA) intended to bring about a fully liberalized air transport market between Europe and North America. To implement this concept will require that EU Member States hand over negotiating authority for an open skies agreement with the U.S. to the European Commission in Brussels. In principle, the EU transportation ministers agreed upon the transfer of authority early in June, 2003. By creating a single EU-U.S. open skies agreement to replace the existing bilateral treaties of the 15 EU Member States would have a very profound impact on future air transport relations between the EU and the U.S.. and other countries and would reshape fundamentally the global aviation industry.

The transformation to a single EU-U.S.. open skies agreement is essentially a question of time but is bound to happen in the not too distant future. In practice, it would mean that any airline in the EU and the U.S. could operate scheduled services to any point on either side of the Atlantic, without needing permission on routes, frequency of flights or tariffs. As a direct consequence, it would also bring about a further consolidation and restructuring of the airline industry throughout the world.

An increasingly interrelated global and free-market economy has brought about an unprecedented focus on international competitiveness and commercial aviation is no exception. During the last two decades, international civil aviation has been subjected to the most profound changes in its history and this will continue in the years to come, coupled with the need to maintain a safe, secure and efficient aviation system worldwide. A potent combination of liberalization of air services, advanced communications technology, globalization of markets, international alliances and privatization of airlines, airports and air traffic control services are the major factors challenging established management and business practices in the future. It is also worth noting that by the year 2020, according to World Bank projections; nine of the world's top 15 economies will be from the Third World of today, including China, India and Indonesia.

There will be a continuous restructuring of the world air transport industry when a limited number of mega carriers will emerge and being operated on strictly commercial principles, to enable them to finance future fleet acquisitions and to raise large amounts of capital, as do other industries. The smaller airlines, especially airlines in developing countries, will need to establish alliances and cooperative arrangements with carefully selected partners in order to ensure their economic viability in the future. More and more governments, in particular the U.S. and the EU favor open skies agreements.

The tragic events of September 11, 2001, and their aftermath, the war in Iraq and the outbreak of SARS early in 2003, contributed to a sharp deterioration in confidence of the traveling public across the globe insofar as air transport was concerned. This, combined with the evidence that the world economy had proved to be weaker than expected in the period before the terrorist attacks, further reduced the prospects for a global recovery of the airline industry in the year 2003. In 2001 and 2002, a total of 1,623 million and 1,615 million passengers, respectively, were carried by the world airlines, compared with 1,656 million passengers in the year 2000. At this stage, it is difficult to predict how quickly the confidence of the traveling public will rebound since much depends on progress made in the war against terrorism and the implementation of an effective and integrated global aviation security system.

As a direct consequence, the traveling public and the world airlines are being faced with a new set of steep security charges, in addition to the very large number of air travel taxes, special fees and surcharges already in effect at present. Today, this covers well over 700 different taxes, fees and charges worldwide, such as an airport improvement tax, a security tax, an arrival and departure tax, a ticket tax, an air navigation charge and insurance and fuel surcharges, not to mention a goods and services tax. It is not uncommon that for short-haul flights 50 percent or more has to be added to the ticket price for taxes, fees, charges, etc. Clearly, this situation is having a very negative impact on the future of short-haul operations and their economic viability. In principle, government security services should be provided free of charge to the traveling public, in the same way as other modes of transport, highways, bridges, school; and other place of work, to mention a few.

The taxation system pertaining to aviation has been around for half a century. When air travel took flight after the Second World War, it was considered to be a luxury way of travel and ever since governments have imposed on airlines and passengers increasing numbers of taxes, fees and surcharges, which seriously accelerated with new security taxes after the terrorist attacks of September, 2001. For example, in the U.S. in the year 2001, domestic air travel produced about USD 65.0 billion in revenue for the U.S. airlines, which had to pay USD 13.4 billion in taxes and fees, representing more than 20 percent of airline revenue. By comparison, the losses of the U.S. airlines were nearly USD 8.0 billion in 2001 and more than USD 9.0 billion in 2002, and a turn-around to profitability is not expected soon.

To adapt to the current economic realities and new technological advances, especially the new threat to civil aviation security, the world airlines through the International Air Transport Association (IATA) and regional airline associations should develop and implement new innovative strategies and programs, with special emphasis on motivated and well-trained staff, to increase productivity and reduce unit operating costs so as to achieve an adequate return on investments. Similarly, the world civil aviation authorities through the International Civil Aviation Organization (ICAO) and the airport administrations through the Airports Council International (ACI) should continue to focus on the development and implementation of progressive policies, and new facilities and systems designed to improve the global civil aviation infrastructure. A major challenge, which will require the collective efforts and creative ideas of all parties currently involved in international civil aviation.

Chapter 8

Aviation Strategy Implementation

> If you do not think about your future, you cannot have one.
>
> John Galsworthy[1]

> Strategy without tactics is the slowest route to victory. Tactics without strategy is the noise before defeat.
>
> Sun Tzu[2]

It has been said that the winners of the future will be those who embrace change by addressing the challenges that confront them at the present moment. However, simply recognizing the challenges that face an industry is not enough. In unstable times, there is always a great deal of discussion about the difficulties of dealing with these challenges. Discussions concerning problem-solving strategies can be endless, but what is required is the ability to implement fundamental changes and fundamental strategies.

Executives are often baffled when they cannot seem to implement their strategies. Imagine the following scenarios:

- Faced with declining profits and tough economic times, top managers of a major carrier jointly brainstorm where they envision the company to be in the next several years. Their discussion centers around how to pull out of difficult times, and they come up with a plan. As the months go by, however, few of their targets are realized and key managers begin to leave the company.
- A multi-billion dollar airline operates for years with a general growth strategy. As the economy tightens and affects the company, the executives refuse to narrow their focus. Ultimately, the company is impacted by the recession much more than their competitors.
- The CEO of a major airline discusses her strategies with the executive committee and then watches in frustration over the next year as all of her major goals fail.
- A new no-frills, low-cost airline enters the market and customers respond favorably. Not to be outdone, a full-service competitor cuts prices between several city pairs to match the prices of the low-cost airline. However, the cuts are too deep for the full-service carrier to sustain.

---

[1] See <http://www.brainyquote.com/quotes/authors/j/john_galsworthy.html>.
[2] See <http://www.brainyquote.com/quotes/authors/s/sun_tzu.html>.

Scenarios like these are seen every day in industries ridden by high-fixed costs. More specifically, the airline industry tends to be a difficult industry in which to provide sustainable long-term profits, especially among the ranks of full-service carriers. Why can some companies implement their strategies successfully and others fail? Strategy is about change, and change is never easy.

Change is especially difficult today in an era of rapid technological developments. With the new global economy of the twenty-first century, the stakes for dealing with change are higher than ever. It is very difficult to institute any kind of change; most companies cannot manage the process and the results are less than satisfying. The fact is that 70 percent of all change initiatives fail.[3] This is an astonishingly high number. What makes it even more alarming is that it does not matter how well a company's management has gone through the strategy formulation process. The management could have been meticulous in every aspect of the analysis, yet this does not guarantee successful implementation. Like the last scenario, sometimes managers forget what they had planned and just react to their competition.

Conversely, some people believe that the reason strategy implementations fail is because companies are too quick to change. They believe that some companies are so eager to transform themselves because they think it is the right thing to do that the companies get immersed in a potpourri of strategic initiatives, which do not seem to work together. In other words, companies are so busy changing for the sake of change that they lose focus on why the change was necessary in the first place. In reality, no matter what the change, a company can come out somewhat battered by the process; it is draining both humanly and economically.

How does management make sure they are successful in implementing their strategy for change? There is no guaranteed prescription for success, but one thing is necessary: communication. Managers have to be willing and able to communicate the reason for the change clearly and convincingly to their employees. If employees do not support the transformation, it probably will not succeed. Why? Because top management cannot implement change successfully all by themselves. It takes people at all ranks to help carry out the strategy. By the same token, it is essential that top management support the strategic change, because change needs a champion, and there is no one better to be the champion than top management. If employees can see the need for the change and gather the enthusiasm for it, they are more likely to join in. When managers can turn the transformation process into a company-wide campaign, with periodic milestones to accomplish, enthusiasm grows.

As noted before, there is no prescription for guaranteed success, no roadmap to follow. The plans are based on what has or has not worked in the past. However, because one strategy has worked well for one manager does not mean it will work well for another. People are different. Some are better at motivating and building

---

[3] M. Beer and N. Nohria, "Cracking the Code of Change," *Harvard Business Review*, 78, no. 3, 2000: 133–139.

an employee-committed environment, and some are better at the nuts and bolts of operations. Strategy implementation has to be based on the given organization and its managers. Like strategy formulation, implementation of the strategy must be done by relying on the strengths of the organization and its managers.

Denver Airport Revisited[4]

Remember the problem with the Denver Airport in Chapter 4? Not only were there environmental problems, there were also implementation problems. For example, there was a major communication breakdown between those involved in building the airport and the airlines that would be using the facilities. Initially, the airlines were not involved in the design of their concourses. As a result, changes had to be made after the fact, causing delays in time and considerable losses in dollars. Likewise, while the airport authority was negotiating a contract for a new and improved baggage system, this system did not meet the needs of all of their airline constituents. Again, this resulted in a loss of time and money and the implementation of more than one baggage system. Ultimately, the poorly thought out implementation process cost the city of Denver millions of dollars and caused resentment and frustration to airport users.

Strategic Leadership: A Key to Successful Implementation

If strategic leadership is not present, it is likely the implementation process will be a failure. Strategic leaders perform four key processes:

- Preparation.
- Leadership.
- Change.
- Partnership.

Preparation

The strategic leader assures that the organization is ready for change, that it is ready to take some steps into the unknown. In doing so, the strategic leader must make sure that all management is behind the process and ready to support the decisions

---

[4] Lynda M. Applegate (et al.), "BAE Automated Systems (A): Denver International Airport Baggage System," *Harvard Business School Case*, Harvard University Press, 1996: 1–15.

that are made. Top management is generally aware of some of the political concerns and issues that might need to be addressed prior to implementation. As a result, sometimes preparatory work must be done to ensure success before a full-fledged implementation can begin.

Leadership

Leadership is about vision and foresight. The strategic leader must facilitate the vision and get everyone internally involved in the vision. At the same time, the strategic leader must remain keenly aware of the public's view of this vision as well as the public's concern with how the vision might affect them.

Change

Strategic leadership is needed to solidify the strategy with managers, employees, and the general public. Making changes means focusing and aligning budgets as required by the strategy. This means that sometimes management has to take away from one departmental budget to supplement another and this is not a popular process, as one might expect. Unless handled appropriately by the strategic leader, change can lead to a great deal of hostility.

Partnership

Sometimes change means working with others outside of the organization. In most cases, strategic performance levels cannot be attained without working with the public as well as other outside organizations. It is up to the strategic leader to facilitate this collaborative effort. First, the leader must facilitate and encourage the organization to work with these outside entities. Likewise, the leader must make sure that there are no roadblocks or, if there are, that these roadblocks can be handled appropriately.

Knowing When to Hold it and When to Outsource

An important aspect of the implementation process is knowing when to hold on to parts of the value chain and knowing when to outsource them. This is a decision a good strategic leader needs to be able to make. Sometimes implementation includes making the organization more flexible. One way of adding flexibility to a company is to reduce some of the staff activities. In other words, there are key components of the value chain that can be outsourced. A strategic leader realizes that all the supporting activities in the value chain are services that in many cases can be done more economically if outsourced. If a company is an expert at something, it can generally perform this activity better and cheaper than if it is only one of the many functions of the business.

Outsourcing Aircraft Maintenance

Outsourcing operational activities historically viewed as core airline activities, such as ground handling, catering services, or maintenance, has been met with controversy, in particular, because of potential labor conflicts from lost jobs. Moreover, concern for quality, continuity and safety have been cited as reasons core airline activities should be retained by the airlines.

One area that airlines agree on, whether LCC or legacy, is that aircraft maintenance is considerably cheaper when performed by outsourcers. JetBlue, Southwest, America West, Northwest, Delta and United are among U.S. carriers who outsource major maintenance of their aircraft to contractors, which in some cases are located outside of the U.S.[5] America West and JetBlue have contracts with maintenance facilities in El Salvador; Southwest has always outsourced its major maintenance; Northwest uses companies in Singapore and Hong Kong to maintain its wide-body fleet; Delta has heavy maintenance contracts with Avborne of Canada and Air Canada Technical Services; and United Airlines also uses outside contractors for heavy maintenance.[6]

According to the traditional model, airlines employ their own mechanics who are licensed by the Federal Aviation Administration (FAA). High labor costs in the U.S. and mechanic scarcity, which contributes to the higher labor costs, may increase the likelihood to outsource maintenance. Some airlines, however, do not believe that maintenance outsourcing fits well with their business model; for example, American Airlines outsources only 20 percent of its maintenance and none of its heavy tear-downs.

The issue of maintenance outsourcing is somewhat controversial because some critics link it to safety concerns. An example which is often cited is ValuJet flight 592, which, in 1999, crashed into the Florida Everglades after taking off from Miami International Airport, killing all 110 passengers and crew on board. The crash was attributed to oxygen canisters improperly stowed in the aircraft's hold by maintenance employees working for an outside contractor. Yet not everyone agrees with the point of view that outsourcing maintenance can cause safety concerns; others assert that maintenance is always subject to high quality standards enforced by regulators in the U.S. and elsewhere.

---

[5] See <http://www.consumeraffairs.com/news04/2005/airline_maintenance.html>.
[6] Ibid.

There are additional reasons why management might want to consider outsourcing. If they outsource the nonessential activities, the company's organizational structure becomes flatter. A flatter structure means less bureaucracy, resulting in greater decision-making speed. This, in turn, helps with the company's competitive responsiveness.[7] Yet this is not to say that a company can outsource activities that are considered to be core competencies or core activities in the value chain. This would create a disconnect for the organization, which can be a problem for companies that decide to outsource. However, outsourcing activities that are secondary to the firm – and outsourcing these activities to world-class companies that specialize in these functions – allows the firm to concentrate on its own core competencies.

There are cases that outsourcing has not been viewed as the right strategy for airlines and the area of reservations serves as a good example of such a case. Large airlines such as United, American and British Airways, to name a few, generate more than half their bookings through outsourcing to global distribution systems, such as Sabre or Galileo. JetBlue, however, one of the most successful carriers, is no longer part of the Sabre global distribution network. JetBlue considers a core competency of the company to be its website sales and distribution system. Its management strives to keep distribution costs down as part of the comprehensive strategic model of linking low costs to low fares. Using Sabre or any other global system would be the same as outsourcing the company's competitive advantage.[8]

Putting Together the Right Staff

Implementing change may involve the recruitment and training of capable individuals to carry out the strategy. It would be naïve to believe that nothing internally has to change. There is always a better way to do things. Obviously, a company must start with a good management team that embodies the vision and values as set by the strategic leader. Sometimes the existing management team is fine; at other times the team only needs a little fine tuning; and still at other times it is necessary to go out and find the right person or persons to lead the change.

This is particularly important in the case of complete strategic turnaround. Sometimes it is important to bring in a "turnaround manager." A turnaround manager is one whose primary job is to make the change happen. In situations like these, it is commonplace to bring in someone or a team of people from the outside. Many times these individuals stay with the organization only until the change is completed. Often, these individuals serve as "the bad guys" when the change process involves the loss or reassignment of jobs for the existing staff.

---

[7] J.B. Quinn, "Strategic Outsourcing Leveraging Knowledge Capabilities," *Sloan Management Review*, 40, no. 4, 1999: 9–21.
[8] See <http://informationweek.com/story/showArticle.jhtml?articleID=55800557>.

It is important to have a team of highly skilled individuals who can play varying roles in the implementation process. These individuals must embody the beliefs and visions of the strategic manager. Without this supporting chemistry, employees can become skeptical, and the implementation process can be compromised.

However, good management is not enough. Company management has to make sure that they have the right employee team in place. Sometimes in an implementation process, management has to move people around to showcase their particular skills. There is nothing wrong with doing this as long as the individual is aware of his or her importance to the overall goal. Most people are willing to get behind a change process if they understand why it is necessary and how they can contribute.

The Case of United Airlines[9]

Glenn Tilton took the helm as CEO of United Airlines just in time to steer the company through the U.S. aviation industry's largest bankruptcy ever. His priorities were as follows:

- Restore employee trust and customer and investor confidence.
- Address near-term financial issues.
- Develop long-term plan for renewed growth.

Tilton, a former oil executive, replaced Jack Creighton, who had been charged with helping to lead the carrier in the wake of September 11. Creighton, who was promoted from within, replaced ousted CEO James Goodwin just a year before, when United was plagued with union turmoil and high costs.

Both Creighton and Goodwin were considered interim executives, helping the airline to emerge from its numerous problems, while Tilton has been viewed as a "rehabilitation outsider." Tilton had a reputation for and experience with helping companies facing serious challenges.

Nevertheless, in today's era of lean manufacturing and service operations, it is important to make sure that the remaining staff has the skill set to move the company forward. It is important, therefore, to recruit and retain the right people for the organization. At times, the right person might not automatically be the straight-A person from the top business or engineering school. Instead, the right person is the person with the skill set necessary to meet the company's needs and the personality and career goals to fit the culture of the organization. While some organizations only look for the top MBAs or undergraduates from the top ten schools in the country,

---

[9] See <http://www.forbes.com/execpicks/newswire/2003/11/24/rtr1159083.html>.

other companies, like Southwest Airlines, look for people that best fit their corporate culture. Southwest Airlines' goal is to hire people who like to have fun on the job, and they work hard to achieve this goal. In their hiring process, Southwest utilizes interviews with customers to make sure that the people they hire in their customer-contact jobs have the necessary personality skills to work with the public. Putting on a good face is important at Southwest, and it is important to the airline management that the employees genuinely show their enthusiasm on the job.[10]

No matter what the company, probably the most important element of the organization is its human capital. Investment in human capital can make or break an organization, particularly in the service industry, and benchmarking can be very instrumental at this point. By looking at leaders in areas like customer service, other companies can study how these leading firms recruit, hire, and retain their employees.

JetBlue's Organizational Behavior[11]

JetBlue, one of the most successful airlines in the U.S., has embraced a strong, harmonized organizational culture, which it reflects in its hiring policies. Five values make up JetBlue's culture: Safety, Caring, Integrity, Fun, and Passion. These five core values permeate the culture of JetBlue and are the basis upon which JetBlue categorizes potential employees. Vincent Stabile, JetBlue's VP of People, stated that he has an equal interest in and focus on people as he does cost. With a turnover rate as low as 10–12 percent, the company attracts and retains the best employees by treating them with respect. Lower turnover, consequently, generates a variety of benefits, including financial ones.

Matching the Right Organizational Structure to a Strategy

Which came first, structure or strategy? This is like asking, which came first, the chicken or the egg? While it may appear that structure comes before strategy, this is not always the best scenario. Many experts believe that in order to build an effective, strategy-supportive company, it is necessary to allow strategy to lead structure. In other words, a company must be prepared to be flexible enough to change the structure to enable the strategy. There are five different patterns of organizational structure.

---

[10] Charles A. O'Reilly and Jeffrey Pfeiffer, "Southwest Airlines: Using Human Resources for Competitive Advantage (A)," *Harvard Business School Case*, Harvard University Press, 1995: 1–24.

[11] See <http://www.jetblue.com/workhere/culture.html>.

The Simple Structure

Most organizations begin from a very simple structure. A simple structure is an organizational form where the owner or the top person (e.g. manager) makes all of the major decisions. In this context, decisions are disseminated from the top. In a simple structure, there is little specialization of tasks, little formalized procedures and really very few, if any, rules of the organization. There is less sophistication in firm infrastructure, including information technology. This is not to say that information technology is not important but rather that it generally has not yet evolved. The owner/manager is involved in the day-to-day operations of the business, sometimes putting in his or her own money to keep the business afloat.

A simple structure is generally characterized by a single product and a single geographic location. Often, there is a great deal of flexibility with the simple structure. Generally, there is less investment in brick and mortar and it is much easier to make rapid changes. Figure 8.1 depicts a simple organizational structure.

```
         ┌─────────────────┐
         │ Owner/Manager   │
         └────────┬────────┘
                  │
                  ▼
         ┌─────────────────┐
         │    Employees    │
         └─────────────────┘
```

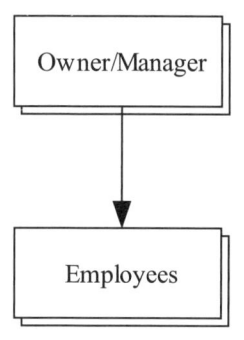

Figure 8.1   Simple Organizational Structure

As evident, there is no formalization to this structure. For a small firm, it is often the best option. However, as a firm begins to grow, some formalized processes are necessary. It becomes obvious that it is impossible for one person to handle all aspects of the company, and there is a need for more complex information systems or different skill sets among management and employees. In other cases it might be time to launch a new product. This is the moment to begin looking at a new organizational structure.

The Functional Structure

When changes are needed in terms of expansion, this is often the time a firm should consider a functional organizational structure. The functional structure is headed by the CEO and has a limited number of individuals in functional areas of responsibilities.

These areas may include human resources, operations (service or manufacturing), R&D (in some cases), finance or accounting and marketing and sales. In this structure, it is important for the CEO to coordinate communications between the functional areas. Like the simple structure, most communications are vertical, primarily because each of the functional areas are self-contained, which is also one of the problems with a functional structure. The individuals heading these areas are likely to have little interest outside their primary domain, tending to look at things for the betterment of their own functional area rather than the good of the company. A good CEO has to manage this process and direct these energies to work for the company.

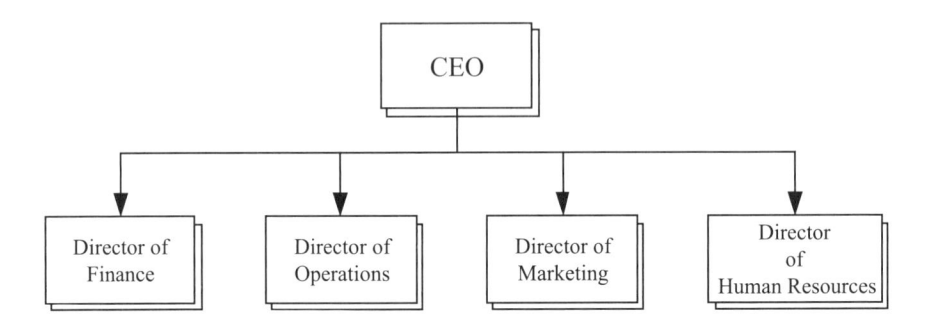

Figure 8.2   Functional Organizational Structure

A typical functional structure is depicted in Figure 8.2. Note that there are different boxes that can be placed under the CEO; however, decisions are still generally initiated from the top.

Southwest Airlines is organized in a "modified" functional structure, as depicted in Figure 8.3. Activities revolve around business processes within the functional hierarchy.

Multidivisional Structure

A multidivisional structure is often introduced into a company when there is a need for greater diversification in the business. Deciding to offer the same product in different markets or deciding to offer multiple products often gives way to a multidivisional structure.

When using a multidivisional structure, the CEO needs some additional help in order to process the growing amount of strategic information necessary to run the business. Alford Chandler explained it quite clearly: "The M-form (multidivisional form) came into being when senior managers operating through existing centralized, functionally departmentalized structures realized they had neither the time nor the

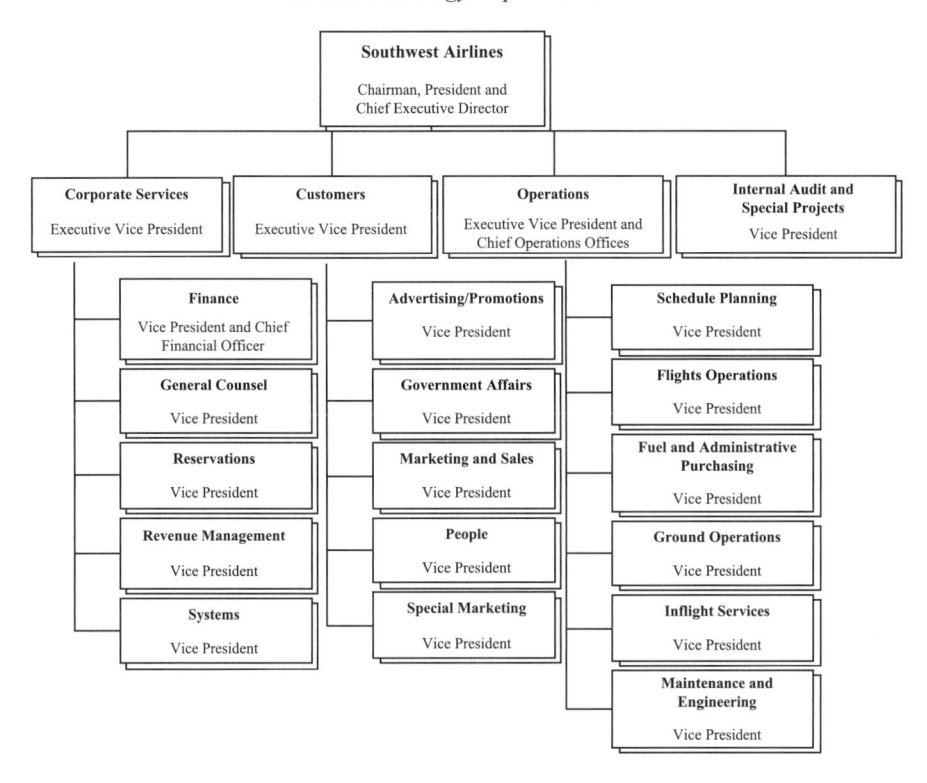

Figure 8.3 **Southwest Airlines' Modified Functional Organizational Structure**

necessary information to coordinate and monitor day-to-day operations, or to devise and implement long-term plans for the various product lines."[12]

The multidivisional structure is characterized by different operating divisions. The operating divisions represent a separate business, a geographic area, or a separate cost/profit center each, of which has responsibilities for day-to-day operations of their unit.

It is important to realize that while most global businesses are organized in a multidivisional structure, this structure does not always enhance a firm's performance nor does it always increase the bottom line. With this type of structure, not only does the CEO lose some control, but also the business becomes more costly to operate. It becomes necessary to have more than one highly paid executive, and the divisional managers must be able to understand the operations, make strategic decisions, and understand how these decisions fit into the overall picture of the organization.

---

[12] A.D. Chandler, "The Functions of the HQ Unit in the Multibusiness Firm," quoted in R.P. Rumelt, D.E. Schendel and D.J. Teece (eds), *Fundamental Issues in Strategy*, Cambridge, MA: Harvard Business School Press, 1994, 327.

154 *Designing and Executing Strategy in Aviation Management*

Moreover, as a company increases in size, it is no longer feasible for the CEO to know the exact contribution of each of the separate product lines in relation to the overall return on investment of the organization. It just gets too complex. Consequently, in order for the organization to operate effectively, the organizational structure must change. Figure 8.4 depicts an example of a geographic structure.

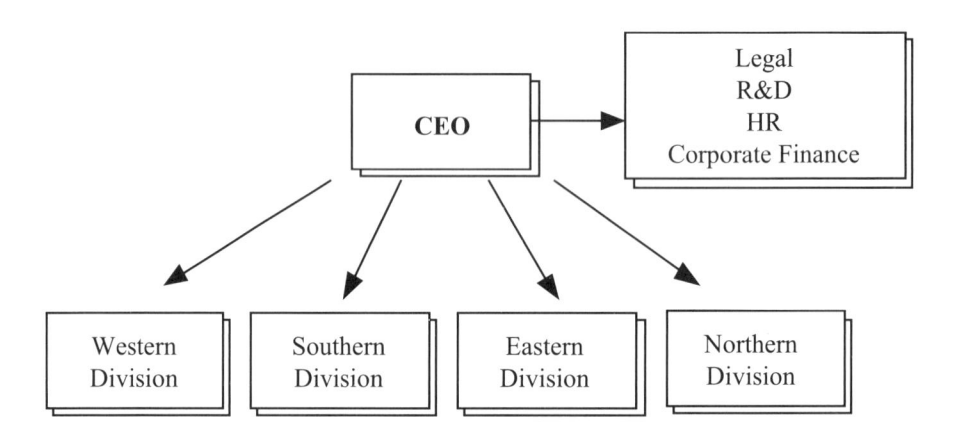

Figure 8.4   Multidivisional Geographic Organizational Structure

Figure 8.5 shows a typical product-line multidivisional structure for an airline. As evident, the airline is divided into two distinctive product areas: low fares and regular/premium fares.

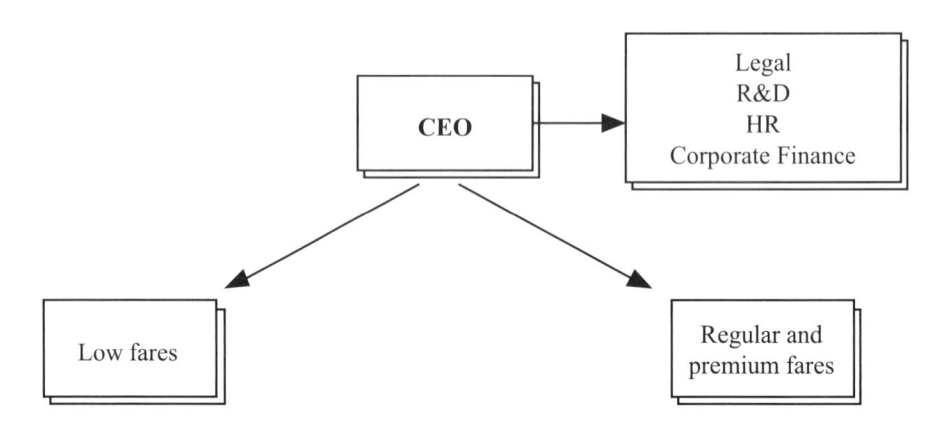

Figure 8.5   Product-line Multidivisional Structure for an Airline

> **Delta and Song**[13]
>
> Delta AirLines provides a good, current example of an organizational structure as viewed in Figure 8.5. In addition to streamlining its fleet, Delta is optimizing its product portfolio to offer the right combination of worldwide service, regional connections, low-fare options, and premium products. This includes launching Song, Delta's low-fare service, and expanding Delta Connection regional airline and Sky Team alliance's advantages. Delta expanded its codeshare alliance with Continental Airlines and Northwest Airlines, in an effort to produce significant revenue benefits when fully implemented, while providing customers with greater travel options. The company is also trying to create sustainable efficiencies for its network by redesigning aircraft turnaround time processes as well as making fundamental changes to its flight staffing and scheduling programs. Delta's Technical Operations facility is streamlining processes and building industry-leading expertise to increase productivity and create opportunities for additional revenue through insourcing. By implementing lean maintenance techniques and utilizing Six Sigma green and black belts to drive operational efficiencies, Delta is enabling flexible skill sets, matching product output to customer needs and optimizing materials and inventory.[14] Delta's operational and product initiatives are expected to bring $1.2 billion in benefits by the end of 2005.

One other form of multidivisional structure is the process form, where the different business processes are separate, stand-alone divisions. In today's virtual world, one can make the argument that companies like Nike, who outsource most of the operations, are organized in a process mode because company officials "manage the process." The same can be said for E-bay as well as Dell Computers. Virtual businesses, where all aspects of the business are outsourced, provide the best description of a process form of multidivisional structure. The practice with some upstart airlines has been very similar, where they function to a large extent as virtual airlines. This practice has been true at the inception of many low-cost carriers, which, in an effort to keep costs down and start flying, have used outsourcing of virtually every service they need, even using, in some cases, AOCs from their aircraft lessors or other airlines. Ryanair started out in this fashion.

---

[13] See http://news.delta.com/print_doc.cfm?article_id=9057>.

[14] Six Sigma is a rigorous and a systematic methodology that utilizes organizational information and statistical analysis to measure and improve a company's operational performance, practices, and systems. This systematic approach attempts to help managers identify and prevent "defects" in manufacturing and service-related processes. There are two levels of certification: green is the basic and black is the master's level. See< http://www.isixsigma.com/dictionary/Six_Sigma-85.htm>.

Strategic Business Units

Strategic Business Units (SBUs) are a more advanced form of a multidivisional structure. The structure consists of at least three different levels. The firm is organized around business portfolios known as strategic business units. The first level, as in all the others, is the CEO. The next level is the strategic business unit, and the final level are the divisions, grouped together based on some form of relatedness. The businesses within each of the SBUs are related to one another, but the SBUs, themselves, are relatively unrelated. Each of the SBUs represents a cost center that reports to the overall corporation.

Most of the functional aspects of the business have corporate operations: human resources, legal finance, strategic planning, marketing, and R&D. Many of these functions are also repeated at the SBU level but report directly to the corporate-level functional offices. The benefit of an SBU is that it allows the company to expand into many unrelated areas, making control much easier. However, the cost of SBUs can be prohibitive. SBUs have been described as individual corporations reporting to one entity. In this regard, the level of sophistication and knowledge at the management level is great and comes with a large price tag with regard to salaries.

Lufthansa's SBUs[15]

After undergoing a turnaround and transformation process from 1992-1994, Lufthansa, a former state enterprise, was converted into a market-oriented and customer-focused company. Now, Lufthansa Group has six business areas: passenger traffic, logistics, MRO, catering, leisure travel, and IT services. More specifically, they also have other companies in the Lufthansa Group, such as the insurer Delvag, Lufthansa Flight Training, and Lufthansa Commercial Holdings.

Lufthansa's Aviation Group has been operating under a philosophy of champion enterprises, which revolves around the concept of SBUs. What this means is that each of the SBUs has as a core strategic objective the attainment of comparative advantage and leadership position in its market segment. Some of the results of this strategy are listed below:

- Lufthansa Technik was able to expand its position significantly in the international MRO market and achieve good results.
- As the smallest strategic business unit in the group, Lufthansa Systems Group has consistently expanded its market position in the airline IT business.

---

[15] See <http://www.lufthansa-financials.de/servlet/PB/menu/1029826_12/index.html?QUERYSTRING=%73%74%72%61%74%65%67%69%63%20%62%75%73%69%6e%65%73%73%20%75%6e%69%74%73>.

Aviation Strategy Implementation

- Lufthansa Cargo suffered from the weak economy that led to a lower demand for air cargo services. By employing a consistent cost management methodology throughout its operations, however, LCAG (Lufthansa Cargo Charter Agency GmbH) was still able to generate good results and maintained its strong position in the market.
- There was progress in the turnaround of LSG Sky Chefs. The company was able to reach its target of a positive segment result.
- At Thomas Cook AG, high priority was given to profitability improvements through strict cost management and an increase in crisis resistance by making the business system even more flexible in a market environment that has become progressively more difficult.

A typical organizational chart for a multidivisional business organized in SBUs is illustrated in Figure 8.6.

The Matrix Organizational Structure

Sometimes the most important part of strategy implementation is flexibility. A company has to be able to work fast and react quickly with a well established plan. Knowing this, some organizations choose a matrix organizational structure. The matrix organizational structure is so named because it creates management responsibility in multiple directions, which defies traditional principles of management. In a matrix organization, there are dual lines of authority and dual lines of budget, providing flexibility and allowing for high powered "brains" to work together to beat the competition.

If a company needs to get to the market fast – beat out the competition – it wants its best and brightest working on the task. A matrix structure allows for this. It brings together the best people to work on a particular project. They complete the project, and then return to their functional areas. Of course, it is more complicated, but it gets the job done. For example, the government might ask for a space shuttle type aircraft to be developed and manufactured. Another government department might ask for a spacecraft to go to Mars. Obviously, these craft would be very different. How could an organization set itself up to complete both projects? It might set up two project groups: the Mars group and the shuttle group. Both would utilize resources, staff, etc. from all of the different departments in the organization. When the projects are completed, these project groups would be disbanded. A typical organizational structure for a matrix organization is illustrated in Figure 8.7.

Figure 8.8 shows more clearly the lines of authority in a matrix organization. Note how employees in this diagram report to their respective vice president as well as to the project manager. Although this structure defies traditional principles of management,

Table 8.1 The Divisions that Make up Lufthansa's SBU Organizational Structures[16]

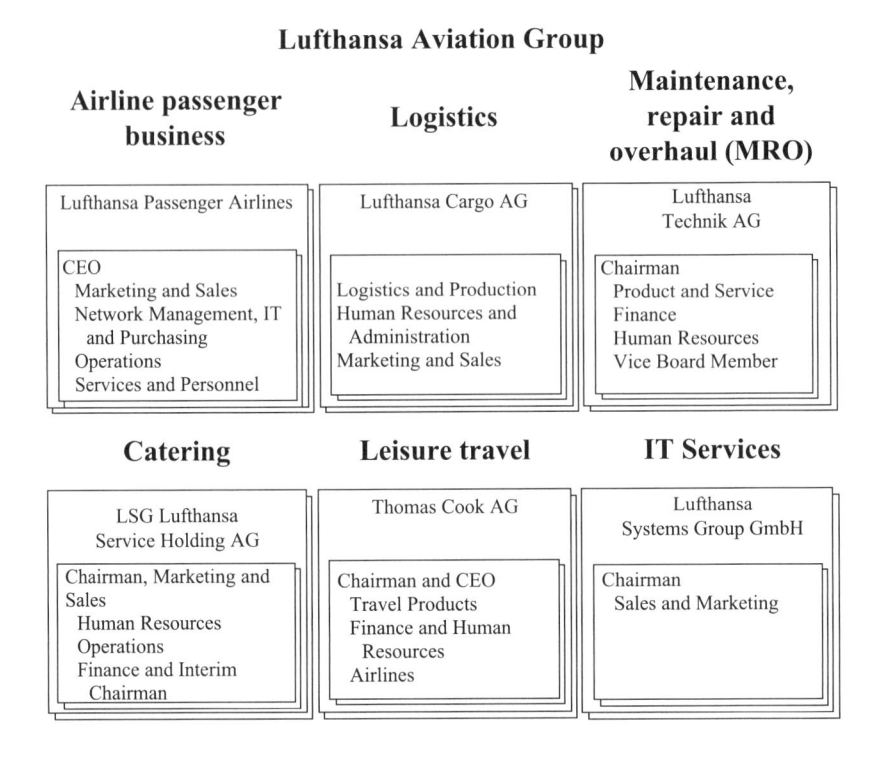

Lufthansa Aviation Group

Airline passenger business	Logistics	Maintenance, repair and overhaul (MRO)
Lufthansa Passenger Airlines CEO Marketing and Sales Network Management, IT and Purchasing Operations Services and Personnel	Lufthansa Cargo AG Logistics and Production Human Resources and Administration Marketing and Sales	Lufthansa Technik AG Chairman Product and Service Finance Human Resources Vice Board Member

Catering	Leisure travel	IT Services
LSG Lufthansa Service Holding AG Chairman, Marketing and Sales Human Resources Operations Finance and Interim Chairman	Thomas Cook AG Chairman and CEO Travel Products Finance and Human Resources Airlines	Lufthansa Systems Group GmbH Chairman Sales and Marketing

which calls for single lines of authority, it provides flexibility to respond quickly to competitive pressures. Matrix organizations have been used quite successfully in the aerospace industry by companies like Boeing.[17]

Boeing: The Matrix Organization[18]

Most of the components within the Boeing organization are arranged in a matrix format, and most employees within Boeing have at least two managers to whom they are responsible. Engineers, for instance, report to a senior executive for

---

[16] Ibid.
[17] See <http://www.boeing.com/news/frontiers/archive/2004/august/i_ca3.html>.
[18] Ibid.

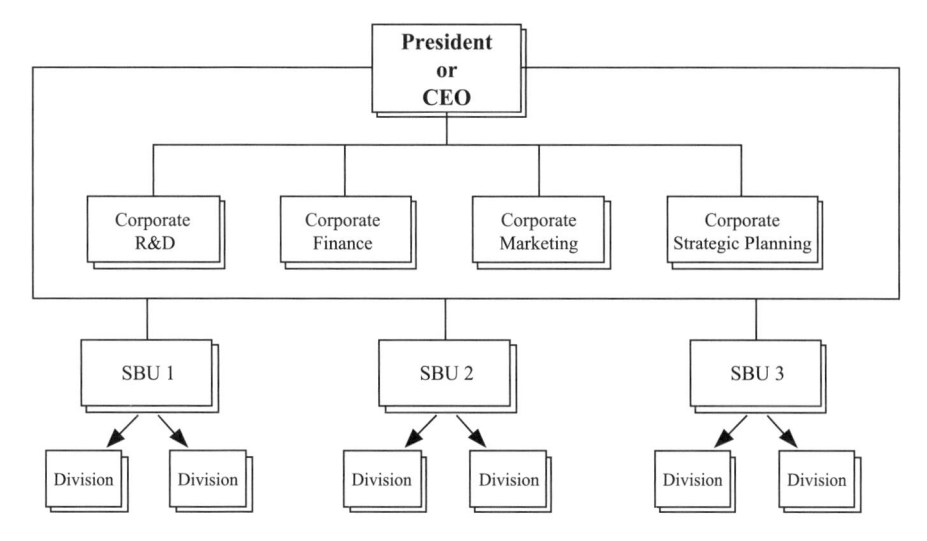

Figure 8.6   Multidivisional Business Organized in SBUs

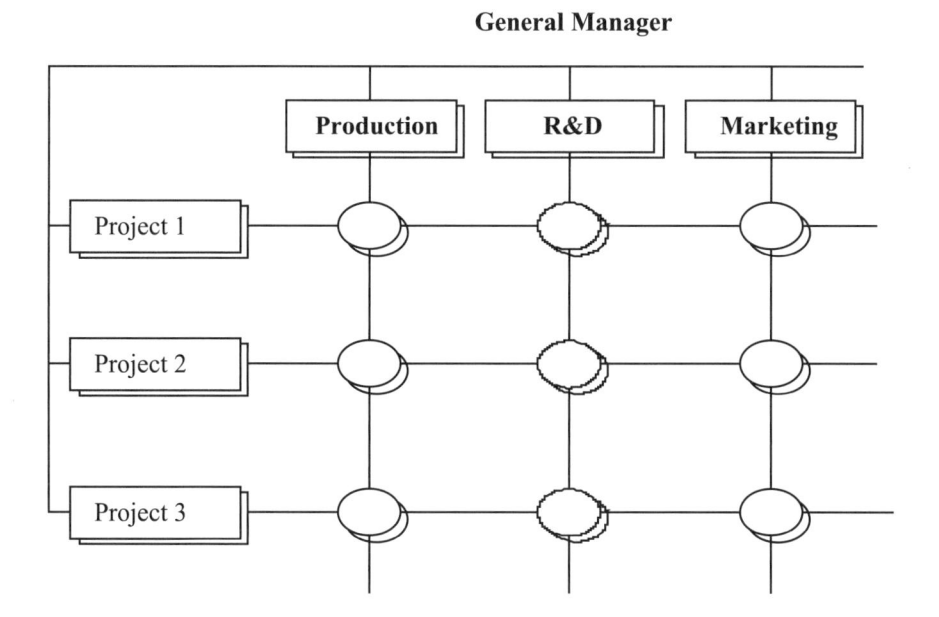

Figure 8.7   Matrix Organizational Structure

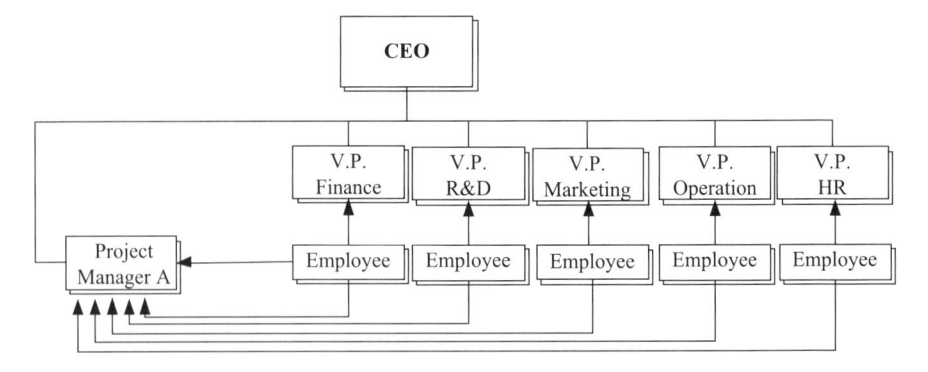

Figure 8.8   Lines of Authority in Matrix Organizational Structure

engineering, and, at the same time, to a specific project team. The project manager has day-to-day responsibilities for the engineer's work but is not responsible for such things as professional training, career tracks, or determination of project assignments. These matters would be the responsibility of the senior engineering executive.

A third dimension to the matrix organization is "site management." Each Boeing facility has a designated site manager, who is responsible for duties such as property management, security, and building maintenance. Although the actual work in each of these functional areas is generally performed by Shared Services Group (SSG) employees, the responsibility for ensuring that a facility runs properly rests with the site manager.[19]

In the Boeing matrix, the functional specialists who are assigned to the business units do not always report on significant matters directly to the heads of their functional groups at world headquarters, that is, to the executives responsible for ensuring uniform practices on a company-wide basis. Correspondingly, the functional executive at world headquarters does not always have direct, day-to-day authority over the individuals in his or her functional organization, who are assigned to the various business units. Although the majority of employees at Boeing have functional reporting responsibility to a functional senior executive, certain employees, such as business development and program management staff, report through management chains to the heads of the major business units.

---

[19]   The Boeing Company Shared Services Group (SSG) provides the company's business units and World Headquarters with common services that support the design and manufacture of aerospace and defense products. These services range from computing resources, telecommunications, e-commerce, and information-management security to transportation, facilities and purchase of non-production goods. The SSG also directs safety, health and

Instituting Total Quality

Organizational theory considers "customer delight" to be a core goal of any organization in its effort to achieve a long-term and sustainable competitive advantage. Organizations should focus on the attainment of high standards and consistent service in an effort to be successful. According to W. Edwards Deming, who considered consumers as the focal point in the quality equation, "the consumer is the most important part of the production line. Quality should be aimed at the needs of the consumer, present and future."[20] This perspective of quality, both in products and services, is determined by what the consumer wants and is willing to pay for. A company's success in implementing its strategic goals is closely tied to the way it competes on the aspect of quality. The lack of both a clearly defined strategic plan and a well-defined implementation plan can be detrimental for an organization in achieving its goals of high quality performance. Organizations that use quality as the cornerstone driver of a competitive strategy have some common links. They have clear strategic goals, vision, or mission that focus on the attainment of customer satisfaction through quality. Difficult to achieve, "stretch goals" are characteristic of companies that tie quality to strategic goals. For example, the implementation of Six Sigma as a means of minimizing defects in companies like Bombardier (manufacturing) or Air Canada (maintenance) is very important to these companies' attainment of their strategic goals of efficiency and competitiveness.

Strategic planning and implementation includes operational plans and policies through which an organization's goals can be achieved. This includes establishing corporate as well as functional goals and keeping employees informed on how these goals should be achieved. In a quality-management program, goals and objectives are likewise established at all managerial levels, and the resources for achieving these goals are provided to managers and employees. Resources need not be monetary and may include employee training and/or the implementation of new or improved processes.

Strategic quality planning typically includes a mechanism for feedback to adjust, update, and correct the original strategic plan. Complacency can be a problem in

---

environmental planning; security and fire services; the hiring, training and motivation of the Boeing workforce; and comprehensive travel services to all Boeing employees. It has companywide responsibility for disaster preparedness response. In addition, Shared Services manages the sale and acquisition of all leased and owned property through the Boeing Realty Company. The role of the SSG is to integrate the operating infrastructure across Boeing, allowing other business units to focus on customer needs and product development and delivery. More information on Boeing's SSG can be found at: <http://www.boeing.com/ssg/ssg-backgrounder.html>.

[20] W. Edwards Deming, quoted in R. Russell and B. Taylor, *Operations Management*, 4d ed. Upper Saddle River, NJ: Prentice Hall, 2003, 615.

this context, particularly if original quality goals are too easily attainable. Therefore, continuously re-evaluating, updating, and revising is very important in keeping abreast of quality goals. Strong corporate leadership is also of utmost importance in the integration of quality into an organization's strategic plan.

In the aviation industry, the applications of total quality can be found in two very different business settings: manufacturing and services. As far as aviation manufacturing is concerned, aerospace and aircraft manufacturers place a very strong emphasis on the manufacturing processes involved in product definition. Product reliability and safety are the two most important strategic goals for aviation manufacturing firms in their quest for new product design. In addition to product definition, aviation manufacturers require marketing, financial, and other business management functions just like all other manufacturing firms. The systems that are used for forecasting product demand, determining material requirements, supply-chain management and optimization and inventory and distribution management are all important parts of operating a manufacturing enterprise, aviation related or not.

Furthermore, while all manufacturers pay attention to planning how they will produce their products, aerospace and aircraft manufacturers also document, for their customers and very importantly for oversight regulatory agencies, how their manufacturing approaches meet all design requirements. Moreover, aircraft and aerospace manufacturers must be able to demonstrate that the product was, in fact, built according to the manufacturing plan, thereby verifying that the design intent for the product was indeed met. Safety is the core concern here.

Information systems are strategically used in manufacturing to support the performance of vital business processes. Aerospace and aircraft manufacturing enterprises have information system requirements that are uncommon to other manufacturing firms. A commodity manufacturer, for example, may acquire an enterprise resource planning (ERP) system that could integrate and help execute a majority of the firm's essential business processes. Aerospace and aircraft manufacturers, on the other hand, must recognize that standardized systems of this nature cannot meet the product definition and manufacturing execution requirements of their particular businesses. In this very specialized environment, final assemblies must usually be matched to a final assembly number, such as the aircraft tail number. While ERP products may support these enterprises' business management and production planning activities, aerospace and aircraft manufacturers need other information technology systems to generate and manage their product design and manufacturing planning. In addition, aerospace and aircraft companies need to control their production processes so that their manufacturing plan and, therefore, their design intent is realized and fully documented.

In the airline industry, the story is rather different and, in a way, much more complicated than in the manufacturing sector. A funded NASA Space Grant and Fellowship Program, conducted by the University of Nebraska at Omaha and Wichita State University, used 15 elements considered important to consumers to develop an airline quality rating. These criteria included baggage handling, customer complaints, denied boardings and on-time arrivals. Table 8.2 below shows the rank order listing for the 2004 Airline Quality Rating (AQR) of the 16 largest U.S. carriers.[21]

As evident, legacy carriers were largely grouped in the mid-rankings, regional carriers dominated the bottom tier, and low-cost carriers topped the chart. The link between quality and competitiveness is more complicated in the airline industry, because consumers use airline travel as an indirect good. This means that customers do not "consume" the airline experience in and of itself, but the airline experience is what brings them to their goal of travel from point A to point B. Sometimes they purchase air travel from travel agents without knowing or caring about the service operator. Some consumers are extremely price sensitive, belong to frequent flier programs, or receive other incentives to fly a specific airline and, therefore, do not take airline rankings into consideration. The point is not that airline quality is of no importance to the consumer but, instead, that it may not be linked as clearly to organizational effectiveness, efficiency and competitiveness as in traditional manufacturing operations.

Table 8.2 2004 Airline Quality Rating

1.	JetBlue
2.	Air Tran
3.	Southwest
4.	United
5.	Alaska
6.	America West
7.	Northwest
8.	American
9.	Continental
10.	ATA
11.	Delta
12.	US Airways
13.	American Eagle
14.	SkyWest
15.	Comair
16.	ASA

---

[21] 2004 Airline Quality Rating (AQR), quoted in *Washington Aviation Summary*, Washington, D.C.: Baker and Hostetler, LLP, 3 May 2005.

Chapter 9

Managing Strategy Execution through Tracking, Support Systems and Controls

> However beautiful the strategy, you should occasionally look at the results.
> Sir Winston Churchill[1]

How does a company determine if it has been successful in its strategic endeavors? When can management make these determinations? It is obvious that these determinations cannot be made unless management knows exactly where they started. In other words, management has to have some kind of control mechanisms in place in order to be able to measure strategic success and the control mechanism has to have a starting point from which to measure.

If properly designed and implemented, the controls in an organization provide for clear insights into how well the company is performing in the marketplace as well as on the bottom line.

Tracking through Information Systems

In the information age, it seems only right that companies have the appropriate information systems in place to enable them to track strategic success. Accurate information is essential to successful strategy and successful implementation. In every aspect of strategic management a company must rely on the availability of information, or how else could management analyze the environment and build strategic scenarios? Consequently, all companies need systems that accurately track and store information. This is commonly referred to as knowledge management. Knowledge management means making sure that management keeps track of what has been done in the past as well as what information is available now.

It is important, however, for a company to think in terms of what is simple with regard to information tracking. This might seem contradictory to everything learned so far. However, having simple systems and processes does not mean being antiquated or outdated. One of the problems often seen in organizational tracking and controls is that management gets wrapped up in developing sophisticated, complex systems involving the deployment of time-consuming software and overly ambitious

---

[1] See <http://www.politicalquotes.blogspot.com/>.

data collection. The result is that flexibility in decision-making is compromised. As noted by Kent Greenes, Senior Vice President and Chief Knowledge Officer at Scientific Applications International, companies too often buy technology to capture information in all forms, creating a repository that lacks the context needed to make the information useful.[2]

The key to good information systems aimed at data collection is to look to simpler tools and create user-friendly systems. It makes no difference how much or how sophisticated the data collection system is if the system is too difficult for anyone to reuse. Raytheon Company has made it a priority to tackle the data repository problem. To access information, the company built a portal using Oracle software to search multiple databases.[3]

Today, there are things like web conferencing, online work spaces, and instant messaging to help people collaborate on projects and link together from different places. Intelligent search engines match people with the necessary information. The goal of all of this is to help people sift through all of the information available, both internal and external to the company, to find the relevant data necessary to track the company's strategic moves. The information might be research data, company financial performance, quality indicators, and competitor's moves and results; there is an endless array of information available. Managers need to be able to monitor, on a daily basis, all aspects of their own operations as well as that of their competitors'.

Internal Systems

As previously noted, continuous monitoring is essential to assuring that the company's strategic decisions remain on track. There are six basic categories of information that are essential.

- Financial performance data.
- Operational data (including quality data).
- Employee data (including turnover).
- Customer data.
- Supplier data.
- Competitor's performance data.

Financial performance data This is something all organizations rely on in order to determine how well they are performing. Financial indicators, such as profitability

---

[2] Tony Kontzer, "The Need to Know: Knowledge Management Has Gone from Pie-in-the-sky Promises to More Realistic Applications," *Information Week*, 18 August 2003, <http://www.informationweek.com/showArticle.jhtml?articleID=13100330>.
[3] Ibid.

ratios, leverage ratios, liquidity ratios, activity ratios and dividend ratios are monitored continuously to assure the company remains on performance targets. For instance, retail companies generate daily sales reports for stores and maintain up-to-the-minute inventory on all items.

Operational data Likewise, operational data is something that is collected daily by most organizations. Production data is collected daily, such as downtime, employee productivity and machine productivity. Included in these reports are often quality and defect rates.

Employee data This includes, as mentioned above, productivity information, but it also involves tracking turnover, absenteeism, and tardiness rates. Some organizations, like Lincoln Electric, rely on piece-rate work where employees are paid by the work performed. This allows the company to keep track of individual employee performance.[4]

Customer data This helps monitor what products or services are being purchased and consumed. This serves as a tracking system and helps companies make alterations before too much time expires.

Supplier data This helps monitor the status of inventories, process orders and invoices, and track shipments.

Competitive data This helps monitor the performance of rival companies. This data can be obtained through trade organizations, stock performance indicators, other forms of public record and simple observation.

All of these control systems help managers to monitor what is happening around them. Continuous monitoring allows managers to keep a watch on early indicators of possible problems and, as the tracking continues, it alerts management as to whether or not corrective actions are necessary or, in some more drastic cases, whether or not to scrap the strategic initiative altogether.

E-Commerce as a Support System

The phenomenal growth of the Internet is changing the way we look at business today. The Internet has emerged as a critical backbone of commerce. The number of Internet users in 2002 grew to about 655 million, growing at a rate of 30.5 percent annually over the past three years, and could represent 50 percent of the world population by

---

[4] D.F. Hastings, "Lincoln Electric's Harsh Lessons from International Expansion," *Harvard Business Review*, 77, no. 3, 1999: 162–178.

2008.[5] The growth of the Internet has given way to the growth of e-commerce. E-commerce refers to business activities that involve online transactions to implement or enhance business processes.[6] Today, most businesses in North America have websites and many of these companies conduct business transactions via the web. In fact, according to UNCTAD, the United Nations Conference on Trade and Development, the value of e-commerce business transactions was around $105 billion in 2002 or 4.6 percent of total world e-commerce sales, and it is predicted to reach $858 billion or 6.7 percent of total world e-commerce sales in 2006.[7] This represents a compounded annual growth rate of 69 percent.

E-commerce has had a profound effect on the way business is conducted and the rules of competition. For example, Amazon.com's expertise in developing the "shopping cart" technique set the standard for online purchasing for other e-commerce websites. Today, the company has evolved into a virtual supermarket, selling everything from music to toys to hardware.[8] Still, it is a challenge for e-commerce ventures to retain competitive advantages because business models or strategies can be copied easily. The lack of distinction between one competitor's position and another's has contributed to the failure of e-commerce firms. Pets.com, for example, tried to compete with other online pet retailers offering similar products and similar services with similar sounding names. Consumers had difficulty telling competitors apart and, instead, searched for better prices or stayed with a familiar retailer. Thus, Pets.com did not develop the customer loyalty that it had anticipated.[9]

E-commerce has allowed companies to conduct business-to-consumer transactions by installing systems to allow for electronic payment of invoices or handling of consumer credit cards. All it takes is the installation of software and systems to handle these transactions. Through e-commerce, a company can install both hardware and software systems to allow for automated order-processing and invoicing of both customers and suppliers. In addition, it handles all accounting functions, materials management, and finished goods inventory. Online tracking of inventories, such as airplanes, has helped to improve communications with customers. E-commerce is advancing the way companies communicate; it provides access to other employees

---

[5] United Nations Conference on Trade and Development, *E-commerce in Developed Countries Continues on Strong Growth Path*, Press Release, 20 November 2003, <http://www.unctad.org/Templates/webflyer.asp?docid=4253&intItemID=1528&lang=1>.

[6] G.P. Schneider and J.T. Perry, *Electronic Commerce Second Annual Edition*, Boston: Course Technology, 2001, 3.

[7] United Nations Conference on Trade and Development, *E-commerce in Developed Countries Continues on Strong Growth Path*, Press Release, 20 November 2003, <http://www.unctad.org/Templates/webflyer.asp?docid=4253&intItemID=1528&lang=1>.

[8] Jeffrey F. Rayport and Bernard J. Jaworski, *Cases in E-Commerce*, Boston: McGraw-Hill, 2002, 36–69.

[9] Ibid.

as well as to needed databases, while, at the same time, connecting employees with customers and suppliers to assure a more seamless business process.[10]

Now, let us consider the impact of e-commerce in the airline industry. Online bookings have reduced travelers' dependence on travel agencies. In fact, the reduction of the dependency on travel sales via travel agencies is quite significant in more ways than one as more and more customers buy their tickets via direct sales channels. Both online sales and sales via telephone have seen significant increases. In Europe, for example, in 2004, 67 percent of the consumers chose direct avenues, internet and telephone, to purchase air travel compared to only 45 percent in 1998.[11] Additionally, the market share of online sales for all types of travel has increased from 0.1 percent in 1998 to 7.3 percent in 2004, with projections of 10.8 percent in 2006. This has resulted in sales via the Internet of $17.6 billion EUR in 2004, whereas in 1998 it was $225,000 EUR. Forecasts predict it will rise to $26.9 billion EUR in 2006.[12]

E-Commerce and the Airlines

When one thinks about airlines and e-commerce, immediately what comes to mind is online ticket purchasing at the airline's website, a web-based travel agent, or search engine (e.g. Expedia, Travelocity and Orbitz). However, e-commerce is also fully established within the air cargo business. Several cargo carriers offer the opportunity for their customers to book directly via the web and their website. For example, for its registered clients, Lufthansa Cargo offers a web-based interface called e-Booking. In addition, they offer a much more convenient way to make bookings via Global Freight Exchange (GF-X). GF-X is a platform to make bookings for cargo shipments directly via the Internet. The advantage to the user of these services is that they have detailed access to shipment schedules from several airlines directly via the Internet browser. As a result, they can decide to make the booking on the airline that best suits their needs. GF-X is supported by both cargo airlines and forwarders, such as Lufthansa Cargo, Air France Cargo, DHL, Cargolux and Panalpina, to name a few.[13]

This example shows that the implementation of e-commerce tools helped both forwarders and shippers to succeed in their strategy to reduce the costs and time for booking, to be more flexible, and to increase customer satisfaction.

---

[10] Ibid.
[11] See <http://www.crt.dk/uk/staff/chm/trends.htm>.
[12] Ibid.
[13] See <http://www.united.com/page/middlepage/0.6823.50550.00html>.

Other Essential Support Systems

Is it possible for an airline to provide customer service without important systems like computerized reservations systems, accurate and expedited baggage systems, and, most importantly, a well-structured and comprehensive routine aircraft maintenance program that adheres to required regulation? Although these are all systems that consumers tend to take for granted, when it comes to the management of an airline, the perspective is different. Most airlines do not take these systems for granted; instead, they are considered essential to the company's livelihood.

Companies like FedEx, UPS, and DHL rely extensively on strong support systems to assure the nationwide and worldwide handling of millions of packages on a daily basis. The systems range from monitoring flight plans and weather patterns in terms of their flight operations to tracking the status of packages from initial processing all the way to delivery. Airlines can choose to develop either their own supporting tools or buy existing software and customize it to their specific needs. All the support systems within the processes have their own purpose, but the systems, as a whole, are interdependent and rely on each other. For example, a handling system is composed of several small systems, but each system contributes to the delivery of goods from point A to B on schedule and in satisfactory condition. Tracking tools, as one part of the handling system, helps the carrier to know at every moment where the freight is. Likewise, these tracking tools provide the customer with the information as to when to expect the cargo to arrive for pick-up. But not only are the handling systems important, the sales processes within the cargo carrier are also very important and need support from other systems. Lufthansa Cargo, for example, has several systems supporting its sales process. For market analysis purposes, reporting tools are used, which help the internal sales controller to communicate with the external sales representatives on their activities. Once market analysis is completed, the sales representative is supported by a CRM-tool which stores all customer information (name, address, activities, opportunity management) needed to serve the customer. Following this, web-based tools are used to place the order. Special planning tools for overbooking and budgeting further support the sales process. The CRM-tool, mentioned above, closes the sales process loop with after-sales support, where claims are handled directly in the system.

These systems include many of the e-business tools discussed above, systems that allow them to support the company strategy of next-day delivery. Packages cannot be delivered "positively on time" in "rain or shine" without these strategy-supportive systems. In addition to the process supporting tools, there are also planning and decision making tools. Fleet scheduling and maintenance tools are planning tools of incredible importance. One main characteristic of all these tools is that they are becoming more and more web-based. This means they are accessible from virtually all over the world with a simple internet browser. No software installation or local maintenance is required. In the airline industry this reduces operating costs significantly. The only disadvantages with web-based software are the need for high bandwidth (which is not necessarily available everywhere, e.g. remote places or lesser-developed countries), as well as security concerns since all the information is transferred via the web.

Employee-level Controls

One of the buzz-words in corporate productivity is employee empowerment. Empowerment means giving employees control of their destiny. On the one hand, giving employees more control over their work is a form of motivation, and many people believe employees will work harder as a result. On the other hand, when does control lead to unnecessary risk? Is it possible to give an employee so much control that he or she places the company in a risky position? Concerned with getting the numbers right, an employee might look at cutting corners, possibly making substitutions that could result in company losses or even lawsuits.

Productivity systems can track the performance of an individual employee or group of employees, but what controls can a company put in place that will assure checks and balances on an employee's work performance? It is possible to put into place strict procedures. This might work in some areas like aircraft maintenance, where any deviation can be catastrophic. For example, a Lufthansa Airbus A320 on a Frankfurt to Paris trip (FRA–CDG) encountered a problem, which was linked, after an ensuing investigation, to a maintenance oversight. The left side stick in the cockpit was repaired by a mechanic performing routine maintenance on the aircraft. The mechanic replaced two cables, but he installed the new cables the wrong way. The consequence was that the flight stick responded the opposite of what it was supposed to, moving the aircraft in the opposite direction intended by the pilot (e.g. left movement instead of right movement). During take-off, due to a right crosswind, the captain attempted to correct the position of the aircraft. However, instead of going back to the right, the airplane went further left, almost touching the ground. The first officer immediately took command of the aircraft and brought it back in stable position. One reason for the error was the high complexity of the documentation for the maintenance of the aircraft. However, another failing was that the pilots did not conduct their pre-fight checks properly.[14]

A similar example, which played out more dramatically, occurred on June 10, 1990, on board a British Airways BAC1-11, departing from Birmingham, England, to Malaga, Spain. As the aircraft was climbing through 17,300 feet, the left windscreen in the cockpit (on the captain's side), which had been replaced the morning prior to the flight, was blown out due to the effects of the cabin pressure overcoming the retention of the securing bolts. It was found that 84 securing bolts, out of 90, were smaller in diameter than what was specified by the aircraft manufacturer. The captain of the

---

[14] For the source in German, see <http://www.bfu-web.de/berichte/01_5x004dfr.pdf>.

aircraft was sucked halfway out of the windscreen aperture and was kept from being sucked out entirely by cabin crew while the co-pilot made an emergency landing at Southampton Airport.

The accident report listed the following as factors that contributed to the loss of the windscreen:[15]

- A safety critical task, not identified as a "Vital Point," was undertaken by one individual who also carried total responsibility for the quality achieved and the installation was not tested until the aircraft was airborne on a passenger carrying flight.
- The Shift Maintenance Manager's potential to achieve quality in the windscreen fitting process was eroded by his inadequate care, poor trade practices, failure to adhere to company standards and use of unsuitable equipment, which were judged symptomatic of a longer term failure by him to observe the promulgated procedures.
- The British Airways local management and Product Samples and Quality Audits had not detected the existence of inadequate standards employed by the Shift Maintenance Manager because they did not monitor directly the working practices of Shift Maintenance Managers.

Both the abovementioned examples show that lack of strict adherence to correct standard operating procedures in technical tasks can have disastrous consequences.

But sometimes placing strict controls on employees is not a life or death matter, and it may be stifling and de-motivating. This is particularly true for those tasks where standard operating procedures are not required and where employee creativity may be much more productive. For instance, addressing customer problems may fall into this category. At times, making the customer happy, even if the solution is outside the realm of standard procedure, is the best solution.

Team-based operations are a type of employee control, which can provide a means of good checks and balances. There is nothing more powerful than the watchful eye of a peer. This is particularly true when reward systems are based on group performance. Team members feel accountable for the success and failure of the entire group and serve as a good monitoring system of individual performance. They are generally intolerant of weak performers or unsafe decisions because they realize the future and welfare of the entire team are at stake.

Designing reward systems that support team-based strategic operations is like implementing another quality control system. Such reward systems also prove that the company is willing to follow through on their word. Implementing reward systems, both monetary and non-monetary based, helps stimulate commitment among

---

[15] Department of Transport, Air Accidents Investigation Branch, *Report*, <http://www.aaib.dft.gov.uk/cms_resources/dft_avsafety_pdf_502702.pdf>.

employees. Linking incentives to group performance gives way to more energetic employees working together, while, at the same time, assuring quality work.

Corporate Governance Controls

It is not only employees who need oversight; high-ranking managers and executives are vulnerable to unethical behavior. In the aftermath of Enron and WorldCom, corporate governance control mechanisms are on the mind of high-ranking officials throughout the world. The resignation of the entire Board of Directors of WorldCom and virtually all senior management, including former CEO Bernard Ebbers, just after the collapse of Enron made corporate America pay closer attention to corporate governance. While the crisis involved a small number of companies, it damaged the trust of the entire corporate world.

Corporate governance controls are a necessity in today's business environment. Effective governance must assure that accountability is applied to everyone and according to proper procedures, from executives to the board and from the board to the shareholders. It must also assure that accountability is exercised effectively. In order to accomplish this, standards must be set and values must be identified and implemented.[16]

As noted by Carolyn Kay Brancato of the Global Corporate Governance Research Center, Enron caused a "sea change in the attention given corporate governance and in how directors are viewed by the public, shareholders, employees and the courts."[17] In 2003, a Conference Board recommended that companies can no longer look upon corporate governance as something that they have had forced upon them from the outside by investors and regulators.

The report suggested that corporate governance and best practices should be based on basic legal requirements that outline the fiduciary role of the director:

- The duty of care to be informed and exercise appropriate diligence in making decisions and to oversee the management of the corporation.
- The duty of loyalty to put the interests of the corporation before those of the individual director.[18]

The Board suggested that by instituting best practices the company would have an internal mechanism to manage and control risk. The decisions made by these boards and the actions that they take can be a great deterrent to unethical behavior as

---

[16] J. Browne, "BP's Browne: Transparency Key to Restoring Trust, Forum: Corporate Governance and Accountability," *The Oil and Gas Journal*, 100, no. 44, 2002: 34.
[17] "Conference Board issues Blueprint for Post-Enron Corporate Governance," *The Conference Board News*, 15 June 2003.
[18] Ibid.

evidenced by Enron and other companies. Corporate boards can no longer sit back in an advisory capacity, waiting for management to come to them. The boards that have been the most effective are those that get involved in setting standards of ethical behavior and providing active oversight of the company's business.

These new rules for corporate governance were supported by the enactment of the Sarbanes-Oxley Act and the requirement changes recommended by the U.S. stock exchanges. These new rulings have provided a rigorous framework of federally mandated controls and reforms with regard to corporate governance. Standard and Poors added even more weight to the issue when they came out with governance-transparency ratings for the S and P 500 and individualized company assessments.[19] In essence, it is a type of boardroom scoreboard.

Corporate governance is an important control mechanism to assure that the strategies of the company are executed correctly. In other words, strict corporate governance provides a mechanism to guarantee that the formulation and implementation of company strategies are in the best interest of all the stakeholders: employees, customers and suppliers. Unfortunately, these mechanisms cannot always guarantee that the decisions are flawless or that they will even work. Corporate controls, however, can guarantee that the decision-makers are using the best information and judgment available to them.[20]

Many airlines have adopted corporate governance within the last few years as an indirect response to the financial scandals cited above. United Airlines, for example, adopted their Corporate Governance Policies at the end of 2003, "which the company believes are critical to maintaining the trust of its investors and other key stakeholders."[21] Likewise in Europe, the Lufthansa Group's corporate governance is firmly established. In December 2002, Lufthansa adopted its own corporate governance statement based on Germany's corporate governance code. The German corporate governance code contains statutory regulations for the governance (composed of governance and supervision) of companies listed in Germany. The governance code's core aim is to make the German Corporate Governance system transparent. "Its [the Code's] purpose is to promote the trust of international and national investors, customers, employees and the general public in the management and supervision of listed German stock corporations."[22] Rights of shareholders are clarified in the Code as well as the responsibilities of the two Boards, required for the management of every corporation. The Management Board is responsible for managing the enterprise, and the Supervisory Board appoints, supervises, and advises the members of the Management Board and is directly involved in decisions of

---

[19] E. Zwirm, "New Model for Governance – from MCI," *CFO.com*, 27 August 2003, <http://www.cfo.com>.
[20] Ibid.
[21] See <http://www.united.com/page/middlepage/0.6823.50550.00html>.
[22] See <http://www.corporate-governance-code.de/eng/kodex/index.html>.

fundamental importance to the enterprise. As a rule, the German corporate governance code is reviewed annually against the background of national and international developments and is adjusted, if necessary. The following statement provides an example of the corporate governance structure for United Airlines and an overview of its philosophy vis-à-vis corporate governance.

United Airlines' Statement on Corporate Governance[23]

UAL and its subsidiaries are committed to sound principles of corporate governance, which the company believes are critical to maintaining the trust of its investors and other key stakeholders. The board of directors of UAL directs the affairs of the corporation and is guided by several important documents that lay the foundation of the company's corporate governance. The first is the Code of Conduct, which describes the ethical and legal responsibilities all company employees are expected to uphold and which provides basic guidelines for conducting business on behalf of UAL and its subsidiaries. The next is the Corporate Governance Guidelines, which outline the duties and responsibilities of the board of directors and which establish general governance policies and practices. Supplementing the Guidelines are the individual charters of the board committees, which articulate the specific charges of those committees. Adopted October 30, 2003.

Strategic Controls and Strategic Change

With the globalization of markets, the pace of change has quickened. Managers have less time today to respond to environmental changes. Strategic control plays a paramount role in how well an organization can adapt and respond to change through the identification and interpretation of change triggers. As discussed in Chapter 4, critical forces in the external environment can mandate responses on the part of businesses. For example, airports had no choice but to respond to September 11 with more intensive security measures. These events are sometimes referred to as change triggers.[24]

In order to trigger either operational adjustments or strategic reorientation, managers must collect data and then interpret the data based on some preexisting standard or belief before responding. This process is a complicated "sense-making"

---

[23] See <http://www.united.com/page/middlepage/0.6823.50550.00html>.
[24] S.D. Julian and E. Scifres, "An Interpretive Perspective on the Role of Strategic Control in Triggering Strategic Change," *Journal of Business Strategies*, 19, no. 2, 2002: 141–160.

activity, where organizational managers attempt to give meaning to complex and often indeterminable data. In other words, managers must use their judgment, intuition, and social information processing to make sense of the environment around them.[25]

Since decision-makers do not always detect either internal or external problems to the organization early enough to implement timely and effective responses, strategic controls can be implemented throughout the strategic management process. Developing a formal process of environmental scanning – a method of routinely collecting and reporting data – at a high level within the company can help managers keep abreast of organizational and environmental red flags and provide a control mechanism early in the strategy formulation stage. However, this must be done at a high level. Often, when scanning is decentralized, personnel at lower levels in the company may not be aware of the relevance of certain information. Individuals at higher levels in the organization are more likely to have the intimate knowledge of the company's strategic posture and are, thus, better suited for interpreting information gathered through such strategic surveillance.[26]

Implementation Controls

Implementation controls are mechanisms put into place to assure that the strategy is on target. This involves monitoring performance through the use of milestones. The purpose of these controls is to help managers to determine whether their strategy is on target or whether to alter the basic direction. Managers identify important milestones and set strategic thresholds, and at each milestone the strategy is evaluated against the strategic thresholds. This evaluative process establishes another control because it forces management to examine current strategic performance against expected performance.[27]

---

[25] Charles R. Wright and Michael R. Manning, "Resourceful Sensemaking in an Administrative Group," *Journal of Management Studies*, 41, no. 4, 2004: 623–643.

[26] K.M. Sutcliffe, "What Executives Notice: Accurate Perceptions in Top Management Teams," *Academy of Management Journal*, 37, 1994: 1360–1378.

[27] R. Muralidharan, "Strategic Control for Fast Moving Markets: Updating the Strategy and Monitoring Performance," *Long Range Planning*, 30, no 1, 1997: 64–73.

Bibliography

2004 Airline Quality Rating (AQR) quoted in *Washington Aviation Summary* (Washington, DC: Baker and Hostetler, LLP), 3 May 2005.

Alazmi, Mutiran and Zairi, Mohamed, 'Knowledge Management Critical Success Factors,' *Total Quality Management and Business Excellence* 14 (2003): 199–204.

American Management Association (AMA), *AMA 2002 Corporate Values Survey* (New York: AMA, 2002).

Amit, R. and Livnat, J., 'A Concept of Conglomerate Diversification,' *Journal of Management* 14, no. 4 (1988): 593–604.

Ansoff, Igor H., *Corporate Strategy: An Analytic Approach to Business Policy for Growth and Expansion* (New York: McGraw-Hill, 1965), 152.

Applegate, Lynda M. et al., 'BAE Automated Systems (A): Denver International Airport Baggage System,' *Harvard Business School Case* (Cambridge, MA: Harvard University Press, 1996): 1–15.

Aurik, Johan C., Jonk, Gillis J. and Willen, Robert E., *Rebuilding the Corporate Genome: Unlocking the Real Value of Your Business* (Hoboken, NJ: John Wiley, 2003), 3–19.

Bahaee, Mahmood S., 'Strategy-Comprehensive Fit and Performance,' *Australian Journal of Management* 17 (1992): 195–215.

Barger, David quoted in Bruce Schoenfeld, 'We're Fed Up!' *Cigar Aficionado* July/August 2002, 59.

Barney, Jay, 'Firm Resources and Sustained Competitive Advantage,' *Journal of Management* 17, no. 1 (1991): 99–120.

Beer, M. and Nohria, N., 'Cracking the Code of Change,' *Harvard Business Review* 78, no. 3 (2000): 133–139.

Beer, Michael, 'People Express Airlines: Rise and Decline,' *Harvard Business School Case* (Cambridge, MA: Harvard University Press, 2003), 1–23.

Black, J. Stewart and Porter, Lyman W., *Management: Meeting New Challenges* (Upper Saddle River, NJ: Prentice Hall, 2000), 256.

Browne, J., 'BP's Browne: Transparency Key to Restoring Trust, Forum: Corporate Governance and Accountability,' *The Oil and Gas Journal* 100, no. 44 (2002): 34.

Chew, Bruce, *The Geometry of Competition* (Cambridge, MA, 2000).

Commercial Aviation Today, 19 May 2005.

'Conference Board issues Blueprint for Post-Enron Corporate Governance,' *The Conference Board News*, 15 June 2003.

Coyne, Kevin P., Hall, Steven J. D., and Clifford, Patricia G., 'Is Your Core Competency a Mirage?' *McKinsey Quarterly* 1 (1997).

Davenport, T.H., Thomas, R.I. and Desouza, K.C., 'Reusing Intellectual Assets,' *Industrial Management* 45 (2003): 12–19.

Deming, W. Edwards quoted in R. Russell and B. Taylor, *Operations Management* 4d ed. (Upper Saddle River, NJ: Prentice Hall, 2003), 615.

Dollinger, Marc, 'The Evolution of Collective Strategies in Fragmented Industries,' *Academy of Management Review* 15, no. 2 (1990): 266–285.

Fitzgerald, Tom, 'Corporate Retrenchment or Corporate Renewal?' *The CEO Refresher* (2001).

Flouris, Triant and Swidler, Steve, 'American Airlines' Takeover of TWA: An Ex-Post Analysis of Financial Market Information,' *Journal of Air Transport Management* 10 (2004): 173–180.

French, Sandy, 'CEO Values Replace Corporate Values,' *Canadian HR Reporter* 8 April 2002, 4.

Ghemawat, Pankaj, *Strategy and the Business Landscape* (Reading, MA: Addison-Wesley: 1999), 116–19.

Giblin, Edward J. and Amuso, Linda E., 'Putting Meaning into Corporate Values,' *Business Forum* 22, no. 1 (1997): 14–19.

Gill, T., 'AerLingus Is to Lose Team,' *Airline Business* 13, no. 3 (1998): 10.

Ginter, Peter, Swayne, Linda and Jack Duncan, W., *Strategic Management of Health Care Organizations* (Malden, MA: Blackwell, 1995), 95.

Groenewege, Adrianus D., *Compendium of International Civil Aviation*, 3d ed. (Montreal, Quebec: IADC, 2003), 97–99.

Hamel, Gary and Prahalad C. K., 'The Core Competence of the Corporation,' *Harvard Business Review* 68, no. 2 (1990): 79–91.

Hastings, D.F., 'Lincoln Electric's Harsh Lessons from International Expansion,' *Harvard Business Review* 77, no. 3 (1999): 162–178.

Hitt, M.A., Hoskisson, R.E., and Kim, H., 'International Diversification: Effects on Innovation and Firm Performance in Product-Diversified Firms,' *Academy of Management Journal* 21 (1997): 767–798.

Hitt, M.A., Duane Ireland, R., and Hoskisson, R.E., *Strategic Management Competitiveness and Globalization*, 4d ed. (Cincinnati: South Western College Publishing, 2001), 349–350.

Jain, S. C., 'Environmental Scanning in U.S. Corporations,' *Long Range Planning* 17 (1984): 66–75.

Jarrar, Yasar and Zairi, Mohamed, 'Future Trends in Benchmarking for Competitive Advantage: A Global Survey,' *Total Quality Management* 12 (2001): 906–13.

Julian, S.D. and Scifres, E., 'An Interpretive Perspective on the Role of Strategic Control in Triggering Strategic Change,' *Journal of Business Strategies* 19, no. 2 (2002): 141–60.

King, P., 'Viad Corp. Sells Dobbs International to SwissAir Parent,' *Nation's Restaurant News* 33 (1999): 18.

Kirby, Scott quoted in Bruce Schoenfeld, 'We're Fed Up!' *Cigar Aficionado*, July/August 2002, 58.
Kontzer, Tony, 'The Need to Know: Knowledge Management Has Gone from Pie-in-the-sky Promises to More Realistic Applications,' *Information Week*, 18 August 2003.
Kotabe, M., 'The Relationship Between Off-Shore Sourcing and Innovativeness of US Multinational Firms: An Empirical Investigation,' *Journal of International Business Studies* 21 (1990): 623–38.
Lainos, Giannis, *Fakelos: Olympiaki Aeroporia* [*File: Olympic Airways*] (Athens: Stachy Press, 1992).
Lewis, D. and Koller, M., 'Airline IT Upgrades, Staff Hit Hard,' *InternetWeek* 11 (2001).
McDermott, Kevin, 'Divestitures a Solid Shareholder Value Option in Tough Times,' *Nashville Business Journal*, 30 May 2003.
Miles, Raymond et al., 'Organizational Strategy, Structure, and Process,' *Academy of Management Review* 3, no. 3 (1978): 546–63.
Muralidharan, R., 'Strategic Control for Fast Moving Markets: Updating the Strategy and Monitoring Performance,' *Long Range Planning* 30, no 1 (1997): 64–73.
O'Reilly, Charles A. and Pfeiffer, Jeffrey, 'Southwest Airlines: Using Human Resources for Competitive Advantage (A),' *Harvard Business School Case* (Cambridge, MA: Harvard University Press, 1995): 1–24.
Oswald, Sharon L. and Boulton, William R., 'Obtaining Industry Control: The Case of the Pharmaceutical Distribution Industry,' *California Management Review* 38 (1993): 138–62.
Park, Choelson, 'The Effects of Prior Performance on the Choice Between Related and Unrelated Acquisitions: Implications for the Performance Consequences of Diversification Strategy,' *Journal of Management Studies* 39, no. 7 (2002): 1003–19.
Partridge, Mike and Perren, Lew, 'Assessing and Enhancing Strategic Capability: A Value-driven Approach,' *Management Accounting* 72 (1994): 28–30.
Porter, Michael E., *Competitive Strategy* (New York: The Free Press, 1980), 201–202.
Porter, Michael E., *Competitive Strategy Techniques for Analyzing Industries and Competitors* (New York: Free Press, 1980), 34–46.
Porter, Michael E., 'What Is Strategy?' *Harvard Business Review* 74, no. 6 (1996): 61–78.
Quinn, J.B., 'Strategic Outsourcing Leveraging Knowledge Capabilities,' *Sloan Management Review* 40, no. 4 (1999): 9–21.
Rarick, Charles and Vitton, John, 'Mission Statements Make Cents,' *Journal of Business Strategy* 16 (1995): 11–13.
Rayport, Jeffrey F. and Jaworski, Bernard J., *Cases in E-Commerce* (Boston: McGraw Hill, 2002), 36–69.

Rumelt, P., Schendel, D.E., and Teece, D.J., eds, *Fundamental Issues in Strategy* (Cambridge, MA: Harvard Business School Press, 1994), 327.

Schneider, G.P. and Perry, J.T., *Electronic Commerce Second Annual Edition* (Boston: Course Technology, 2001), 3.

Schoenfeld, Bruce, 'Grounded In Service – Hotels Have Been Successful at Making Travelers Feel Good about Brand Loyalty,' *Cigar Aficionado* July/August 2002, 66.

Securities and Exchange Commission, *American Airlines' Form 10–K405 Annual Report* (2000).

Selton, Rheinhard, 'Factors of Experimentally Observed Bounded Rationality,' *European Economics Review* 42 (1998): 414.

Soloman, Robert, Competitive Intelligence: Scanning the Global Environment (London: Economica Ltd, 1999), 1.

Sutcliffe, K. M., 'What Executives Notice: Accurate Perceptions in Top Management Teams,' *Academy of Management Journal* 37 (1994): 1360–78.

Teisberg, Elizabeth O., 'Strategic Response to Uncertainty,' *Harvard Business School* #9–391–192 (Cambridge, MA: Harvard University Press, 1993), 1–12.

Tully, Shawn, 'The Airlines' New Deal: It's Not Enough,' *Fortune Magazine* 28 April 2003, 79–81.

United Nations Conference on Trade and Development, *E-commerce in Developed Countries Continues on Strong Growth Path*, Press Release, 20 November 2003.

van de Vliet, A., 'Back from the Brink,' *Management Today* January (1998): 36–40.

Wallace, James, 'Aerospace Digest: Low Fares to Europe May be Coming,' *Seattle Post*, 18 February 2004.

Webster, James L., William E. Reif, and Jeffery S. Backer, 'The Manager's Guide to Strategic Planning Tools and Techniques,' *Planning Review* 17, no. 6 (1998): 4–13.

Westley, F. and Mintzberg, H., 'Visionary Leadership and Strategic Management,' *Strategic Management Journal*, 10 (1989): 17–32.

Williams-Tracy, Laura, 'US Airways Needs Hub Here for a Turnaround,' *Charlotte Business Journal*, 6 January, 2003.

Wright, Charles R. and Manning, Michael R., 'Resourceful Sensemaking in an Administrative Group,' *Journal of Management Studies* 41, no. 4 (2004): 623–43.

Zwirm, E., 'New Model for Governance—from MCI,' *CFO.com* 27 August 2003.

Index

ACI 2, 142
Air Canada 15, 41, 68, 101, 103, 110, 147, 161
Air France 29, 30, 104, 109, 110, 112, 169
ALPA 119
American Airlines (AMR) 11, 20, 41, 49, 51, 54, 70, 83, 90, 101, 104, 108, 110, 117, 119, 120, 121, 122, 123, 147, 148, 163
American Eagle 121, 123, 163
America West 20, 147, 163
assessing environmental change 71, 72
ATA 102, 163
ATC/5 131

barriers to entry 37, 57, 58, 137
benchmarking 90, 91, 130, 150
Boeing 3, 47, 89, 99, 118, 123, 158, 160, 161
British Airways 22, 23, 29, 30, 41, 110, 148, 171, 172

CASM 67
Chalk Ocean Airways 20, 30, 31
codesharing 63, 109, 110, 134
competitive advantage 19, 28, 34, 35, 36, 38, 39, 40, 42, 43, 44, 45, 47, 63, 64, 69, 80, 81, 89, 90, 103, 118, 126, 148, 150, 161, 168
competitive strategy 3, 15, 19, 24, 37, 38, 44, 45, 56, 117, 161
corporate governance controls 173
core competency 32, 36, 42, 148
concentration growth strategies 105
Continental Airlines 12, 16, 31, 41, 45, 46, 47, 76, 85, 95, 108, 109, 110, 119, 155, 163
Continental Lite 31, 45, 46
corporate values 5, 6, 7, 8, 10, 15, 16
Corporate vision 5, 6, 8, 10, 127
CRM 86, 87, 170
CRS 86, 140

Delphi Technique 73
Delta AirLines 20, 41, 51, 70, 102, 103, 108, 110, 111, 117, 147, 155, 163
diversification growth strategies 105
  concentric diversification 107
  horizontal diversification 107
  related diversification 106, 107, 108
  unrelated diversification 108
DHL 169, 170
DOT 17

e-commerce 86, 87, 93, 160, 167, 168, 169
employee-level controls 171
environmental analysis 13, 62, 70, 71, 74, 75
environmental scanning 6, 70, 71, 73, 176
environmental uncertainty 74
ERP 162
EU 140, 141
EasyJet 5, 41

FAA 147
FedEx 170
financial objectives 14, 15
first-mover advantage 118
forecasting (environmental change) 70, 71, 87, 133, 162
formulating strategy under uncertainty 132
fractional ownership 33, 34

GATS 138
generic strategy 29, 44
growth strategies 101, 102
  market development 101, 102
  market penetration 101, 102
  product development 101, 102

IATA 2, 40, 109, 139, 142
IBM 54
ICAO 2, 131, 137, 138, 142
implementation controls 176
instituting total quality 161

integrative growth strategies 103
  horizontal integration 103, 104, 115
  vertical integration 103, 104, 105
international strategies 116

Jet Airways 129, 130
JetBlue 4, 5, 20, 26, 27, 41, 66, 95, 96, 97, 102, 103, 147, 148, 150, 163

KLM 104, 109, 110, 112
knowledge management 1, 94, 126, 165

LCAG 157
Lufthansa 100, 110, 117, 156, 157, 158, 169, 170, 171, 174
LSG 157

macro external environment 52
micro external environment 55
Michael Porter 15, 19, 34, 36, 37, 57, 80, 118
Michael Porter's Five Competitive Forces 56, 57, 62, 65
Miles and Snow Typology 34, 36, 37
mission statement 11, 12, 13, 15, 16, 17, 129
monitoring 5, 6, 70, 71, 166, 167, 170, 172, 176
multi-business strategies 99, 100
MRO 86, 156

NASA 162
NetJets 33, 34
no growth strategies 111
  divestiture 114
  liquidation 113
  retrenchment and turnaround 111, 112, 113
Northwest Airlines 109, 110, 119, 155

Olympic Airlines 53
Oneworld alliance 110
organizational brainstorming 74
organizational structures 35, 48, 131, 148, 150, 151, 152, 153, 154, 155, 158, 159, 160
  simple structure 151
  multidivisional structure 151
  matrix organizational structure 157

outsourcing 25, 85, 91, 92, 117, 147, 148, 155

potential entrants 22, 27, 32
power of the buyer market 20, 57, 59, 60
power of substitute products 26, 57, 60, 128, 133
power of the supplier market 57, 58, 59
PSA 104

resource-based view 89
rival competitors 22, 27, 32
Ryanair 5, 20, 22, 23, 25, 41, 155

Sabre 54, 148
SARS 121, 141
SAS 22, 23, 110
SBU (Strategic Business Unit) 156, 157, 158, 159
scenario planning 74, 75
single-business strategies 99, 100, 105
SkyEurope 63, 64
SkyTeam 155
Song 41, 102, 103, 155
Southwest Airlines 3, 4, 5, 20, 31, 37, 41, 45, 46, 47, 48, 49, 63, 66, 83, 84, 85, 89, 95, 131, 147, 150, 152, 153, 163
STAR Alliance 100, 110
strategic cost analysis 80, 88
strategic change 144, 175
strategic controls 175, 176
strategic leadership 145, 146
strategic objectives 15, 99, 114
strategic positioning 19, 49
strategy
  low-cost leadership strategy 15, 20, 21, 22, 24, 25, 26, 28, 31, 34, 36, 37, 49
  niche strategy 15, 20, 29, 30, 31, 32, 33, 74, 117
  differentiation strategy 15, 20, 24, 25, 26, 27, 28, 29, 34, 44, 118
substitutes 22, 27, 32, 133
suppliers 22, 27, 36, 59, 83, 85, 88, 104, 116, 168, 169, 174
sustainability 14, 21, 36, 37, 44, 46, 89
SWOT analysis 65, 125, 127, 129, 130, 132

tracking systems 86
   competitor's performance data 166
   customer data 166
   employee data (including turnover) 166
   financial performance data 166
   internal systems 166
   operational data (including quality data) 166
   supplier data 166

TWA 40, 104, 119, 120, 121, 122, 123

UNCTAD 168
United Airlines 20, 51, 83, 99, 103, 104, 108, 110, 111, 120, 147, 148, 149, 163, 174, 175

UPS 107, 170
US Airways 15, 70, 104, 110, 112, 163

value chain 42, 80, 81, 82, 83, 85, 86, 88, 91, 111, 112, 126, 127, 134, 135, 136, 146, 148
   inbound logistics 82
   operations 82
   distribution and outbound logistics 82
   sales and marketing 82
   service 82
vision 6, 7, 8, 9, 10, 13, 15, 16, 99, 100, 112, 127, 133, 146, 148, 149, 161

WTO 138